SpringerBriefs in Fire

Series editor

James A. Milke, University of Maryland, College Park, MD, USA

For further volumes:
http://www.springer.com/series/10476

Michael S. Hand · Krista M. Gebert
Jingjing Liang · David E. Calkin
Matthew P. Thompson · Mo Zhou

Economics of Wildfire Management

The Development and Application of Suppression Expenditure Models

 Springer

Michael S. Hand
David E. Calkin
Matthew P. Thompson
Rocky Mountain Research Station
USDA Forest Service
Missoula, MT
USA

Jingjing Liang
Mo Zhou
West Virginia University
Morgantown, WV
USA

Krista M. Gebert
Northern Region
USDA Forest Service
Missoula, MT
USA

ISSN 2193-6595   ISSN 2193-6609   (electronic)
ISBN 978-1-4939-0577-5   ISBN 978-1-4939-0578-2   (eBook)
DOI 10.1007/978-1-4939-0578-2
Springer New York Heidelberg Dordrecht London

Library of Congress Control Number: 2014931498

Printed on acid-free paper

Springer is part of Springer Science+Business Media (www.springer.com)

Acknowledgments

The authors wish to acknowledge Jessica Haas, Jon Rieck, and Amy Steinke for data development and support, and Julie Gilbertson-Day and Jeff Kaiden for technical editing support. Nicole Vaillant, Jessica Haas, and Keith Stockman all contributed to the Deschutes case study described in Chap. 5.

The views expressed here are the authors' and do not necessarily represent the views of the United States Department of Agriculture.

Contents

Acronyms

C+NVC	Cost Plus Net Value Change
CFLRP	Collaborative Forest Landscape Restoration Program
CPL	Cost Plus Loss
DCFP	Deschutes Collaborative Forest Project
DOI	Department of the Interior
EC	Existing Conditions
EPH	Expenditures per Hectare
ERC	Energy Release Component
FPA	Fire Program Analysis
FSIM	Fire SIMulation System
GAO	Government Accountability Office
HP	Hierarchical Partitioning
NFDRS	National Fire Danger Rating System
NIFC	National Interagency Fire Center
NIFMID	National Interagency Fire Management Integrated Database
NOAA	National Oceanic and Atmospheric Administration
OMB	Office of Management and Budget
PT	Post-treatment
R-CAT	Risk and Cost Analysis Tool
RMRS	Rocky Mountain Research Station
SCI	Stratified Cost Index
USDA	United States Department of Agriculture
USFS	United States Forest Service
WFDSS	Wildland Fire Decision Support System
WUI	Wildland–Urban Interface

Chapter 1
Introduction: A New Look at Wildfire Management Expenditures

Abstract In the United States, increased wildland fire activity over the last 15 years has resulted in increased pressure to balance the cost, benefits, and risks of wildfire management. Amid increased public scrutiny and a highly variable wildland fire environment, a substantial body of research has developed to study factors affecting the cost-effectiveness of wildfire management activities. This book examines the state-of-the-art in the economics of wildfire management. The introductory chapter presents the broad goal of the book: to take stock of research to-date on the economics of wildfire management and examine a way forward for answering remaining research questions. Subsequent chapters review existing research, present new empirical analyses of fire management expenditures, and examine potential applications of expenditure models for decision making.

Keywords Wildfire management · Suppression expenditures · USDA forest service · Expenditure models · Decision support

The federal land management agencies of the United States [the USDA Forest Service (USFS) and the Department of Interior (DOI)] are struggling to deal with a changing wildland fire environment. Over the past two decades, both the magnitude and the variability of the area burned by wildfire have increased substantially (Calkin et al. 2005; Gebert et al. 2007; Westerling et al. 2006) with the causal factors attributed to past suppression efforts (Arno and Brown 1991), drought conditions (Collins et al. 2006; Crimmins and Comrie 2004; Gedalof et al. 2005; Westerling et al. 2002, 2003), and climate change (Flannigan et al. 2000; Westerling et al. 2006). Rapid population growth in the wildland-urban interface creates further suppression challenges (Cardille et al. 2001; Gill and Stephens 2009; Mozumder et al. 2009). Coincident with these trends, 10-year average Federal suppression expenditures have increased from about $600 million a decade ago (1990 through 1999) to about $1.4 billion over the past decade (2003–2012) (inflated to constant 2012 dollars). As a result, scrutiny of wildland fire management by oversight agencies such as the Office of Management and Budget (OMB) and the Government Accountability Office (GAO) has intensified.

M. S. Hand et al., *Economics of Wildfire Management*, SpringerBriefs in Fire,
DOI: 10.1007/978-1-4939-0578-2_1, © The Author(s) 2014

Because large fires are responsible for the bulk of fire suppression expenditures (USDA Forest Service et al. 2003), interest in understanding the factors that influence expenditures on large wildfires increased along with wildland fire expenditures. Initial research to develop statistical models to either predict fire expenditures or investigate causal factors of expenditures occurred in the 1980s and 1990s (e.g., Gonzalez-Caban 1984; Steele and Stier 1998). But much of the early research into fire expenditures was primarily focused on determining the optimal level for suppression once the pre-suppression budget (e.g., procuring and maintaining equipment) had been determined [e.g., the National Fire Management Analysis System (Lundgren 1999)] and the development of analytical tools for planning efficient fire program investments [e.g., the Fire Economics Evaluation System (Mills and Bratten 1982) and FIREPRO (Botti 1999)]. Research was also conducted into the pre-positioning of suppression resources to minimize damage from wildfires and constrain suppression expenditures (e.g., Fried et al. 2006).

However, as wildfire suppression expenditures began to take up a larger share of the budget of land management agencies, pressure was put on the agencies by Congress, OMB, and GAO to contain expenditures. After the 2000 fire season (the first year the USFS spent over a billion dollars on suppression expenditures) and the subsequent passage of the National Fire Plan, research into the costs of large wildfires became more prevalent.[1]

The goal of this book is to take stock of research to date on the economics of wildfire management and examine a way forward for answering remaining research questions. Of particular interest are the following broad questions:

- What have we learned thus far about the costs of managing wildfires in the United States, and how has this knowledge been used in a management and policy context?
- What gaps remain in our knowledge about wildfire management costs and expenditures in particular?
- How do we answer the persistent and difficult questions about the economics of wildfire management that may help managers improve the effectiveness and efficiency of management efforts in the future?

To answer these questions, the subsequent chapters review a broad literature on wildfire management and extend existing empirical approaches for modeling the determinants and consequences of wildfire management activities. The book is focused on management costs, models of wildfire suppression expenditures, and the application of empirical results to decision support and program planning

---

[1] In this volume the term "expenditures" refers to direct monetary outlays for the purposes of managing wildfire incidents. "Costs" refer more broadly to the negative impacts associated with fire management, which may include direct expenditures on incident management. A full accounting of costs would include a number of non-expenditure impacts that are outside of the scope of this book (see, for example, Butry et al. 2001; Kochi et al. 2012). In most cases this book will refer to "expenditures" unless the broader term is warranted or doing so would create confusion with literature referenced in the text.

efforts. In keeping with the vast majority of past research, the scope of the reviews and analysis is limited to the Federal wildfire management program. Many of the insights discussed in this volume may be applicable to fire management efforts at State and local levels, although in many cases the data necessary to make these connections is not available. Nonetheless, focusing on Federal wildfire management, and in particular on the experience of the USFS, provides a rich description of what is known and not known about wildfire management in the United States.

The outline of the book progresses from a broad survey of existing literature, to extensions of empirical models of wildfire management expenditures, to potential applications of expenditure models in a management planning context. The next chapter provides a broad overview of how empirical models of wildfire management expenditures have been developed and the insights these models are able to provide. Despite a relatively recent history of development of expenditure models, their use has become widespread in the Federal wildfire management arena. The chapter then describes the various uses of these models in decision support, performance evaluation, and land management planning.

Chapter 3 takes an econometric approach to examining trends in wildfire suppression expenditures over time and spatial relationships in aggregate expenditures among geographic regions. Micro-level regression models are then employed to examine differences over time and between regions in the factors that are related to expenditures on individual fires. Taken together, the aggregate and micro analyses can describe overall trends in expenditures on wildfire suppression and how the management of individual wildfires fits into broader trends.

Chapter 4 extends the empirical analysis of a spatially descriptive expenditure model originally presented in Liang et al. (2008). Results from Liang et al. (2008) indicate that using spatial data based on the entire burned area of a fire, rather than data describing the ignition point, can yield a parsimonious expenditure model that accounts for spatial relationships between fire observations. However, data were only available for fires in the Northern Region; in Chap. 4, the model is expanded to include data from the entire western United States for fires from fiscal years 2006 through 2011. This expanded regression analysis can indicate whether the insights from Liang et al. (2008) can be generalized to other regions and whether spatially descriptive data holds promise for improving expenditure models.

The potential for applying regression expenditure models to the analysis of land management planning is demonstrated in Chap. 5. Using an example of landscape-level fuel treatments in the Deschutes National Forest, suppression expenditures are predicted for simulated wildfires that burn in a treated versus untreated landscape. The use of the regression expenditure model allows managers to gauge how a fuel treatment program might affect suppression expenditures in the future and weigh those effects along with the costs and benefits of treatments.

Finally, Chap. 6 summarizes the main findings of the book and points a way forward for future research on topics related to the economics of wildfire management and management expenditures in particular. This chapter highlights some of the remaining and persistent gaps in knowledge about wildfire management expenditures and suggests some research directions for closing those gaps.

As with many aspects of wildfires, much remains to be discovered about the costs of managing wildfires and how programs and policies can be designed and implemented to improve the efficiency of management efforts. Yet remarkable progress has been made in the relatively few years since wildfire suppression expenditures have become a more pressing concern for public land management agencies. This book aims to recognize these achievements and serve as a guide for the next phase of research on the economics wildfire management.

Chapter 2
Development and Application of Wildland Fire Expenditures Models

Abstract Models of fire management expenditures can play a crucial role in the management of wildland fire incidents. This chapter reviews the development and uses of expenditure models, such as the Stratified Cost Index (SCI). Expenditure models are used in decision support tools, budget planning tools, post-season incident reviews, and land management planning. Fire expenditure models are also useful for examining the decisions made by managers on fire incidents and factors influencing suppression activities. Increased exposure to the effects of wildfire and escalating suppression expenditures, particularly in the western United States, suggests that there is a need for improved expenditure models in the future.

Keywords Wildfire management · Suppression expenditures · USDA forest service · Department of interior · Stratified cost index · WFDSS · FPA · Fuel treatments · Management strategy · External factors

2.1 Introduction

Models of wildfire suppression expenditures can play a vital role in managing the costs of wildfire management. First, knowledge of the characteristics of large fires that influence expenditures can be used to predict suppression expenditures for pre-fire budgetary planning (such as the Fire Program Analysis System (FPA) discussed in Sect. 2.3.1 and), evaluating possible suppression expenditure savings due to fuel treatments (discussed in Sect. 2.4 and Chap. 5), or as a performance measure (discussed in Sect. 2.2). Such predictions can also be used in real-time to provide knowledge of which fire ignitions are likely to become large and costly, providing managers with information that could influence the strategies or tactics used on the fire. For example, managers can identify how current suppression expenditures on an ongoing fire compare with similar fires in the past using an application in the Wildland Fire Decision Support System (WFDSS) (described in Sect. 2.3.2). Also, knowing the factors that influence fire expenditures can provide

M. S. Hand et al., *Economics of Wildfire Management*, SpringerBriefs in Fire,
DOI: 10.1007/978-1-4939-0578-2_2, © The Author(s) 2014

insight into which factors may or may not be able to be influenced by management decisions (Sect. 2.5 highlights some of these studies).

This chapter focuses on the development of research on the costs of large wildfires, highlighting the development and use of the Stratified Cost Index (SCI), with some discussion of other suppression expenditure models. Limitations of the SCI are discussed, as well as how the SCI has evolved over time and possible advancements in modeling wildfire expenditures in the future. Finally, the limitations of the current models and emerging developments in expenditure modeling are discussed in the concluding section.

2.2 The Stratified Cost Index

In 1999, the Rocky Mountain Research Station began to explore the possibility of developing models to predict expenditures on individual large wildland fires. In a two-year effort, researchers finally gathered expenditures and fire characteristic data for 218 large (greater than 121 hectares) wildfires reported in the Forest Service's fire occurrence database (National Interagency Fire Management Integrated Database, NIFMID) for the years 1996–1998. Data gathering was hampered by a number of problems. First, matching expenditures (found in the accounting databases of the federal agencies) with fire characteristic information in NIFMID was extremely time consuming and fraught with error as there was no common identifying field between the two data sources. Second, one of the goals of the study was to obtain all of the expenditures for each fire, federal as well as state and local expenditures. For many large fires, several governmental entities can provide resources and incur charges when fighting a wildfire; therefore, to obtain the total suppression expenditures for the incidents in the dataset, it was necessary to obtain expenditure information from each agency involved. However, with the exception of the USFS and the Bureau of Land Management, the accounting systems for the other federal land management agencies within the DOI made it very difficult to collect suppression expenditure information at the fire level. As for state and local expenditures on individual wildfires, most were not available electronically but were contained in paper records kept in boxes, often in the basement of office buildings. Preliminary results from this study showed promise but lack of confidence in the quality of the data collected and the time-consuming nature of the data collection process caused the research to stall.

However, as wildland expenditures continued to climb, pressure was put on the land management agencies to contain expenditures associated with wildfire suppression. The Government Performance and Review Act of 1993, as well as ongoing efforts of the President's Management Agenda, required that Federal Programs develop and report outcome-based performance measures. Accordingly, Conference Report on HR 4818, Consolidated Appropriations Act, 2005 required the Secretaries (Department of Interior and Agriculture) to promptly establish appropriate performance measures for wildland fire suppression and develop a report on performance measures planned for implementation in fiscal year 2006 to be used on an inter-agency basis. Included in this language was the requirement that the agencies develop a measure to report the percentage of fires, using a statistically representative sample, not contained in initial attack that exceed a "stratified fire cost index" (SCI). This index was originally specified as expenditures per acre over energy release component. After discussions between USFS Fire and Aviation Management and researchers at the Rocky Mountain Research Station (RMRS), it was decided the SCI would assess a variety of factors that influence suppression expenditures, rather than focusing solely on energy release component. In fact, the "Stratified Cost Index", as it came to be called, was simply the redevelopment of the regression models first tested by the RMRS in the 1999 study.

For the new effort, data were collected on fires reported in NIFMID for fiscal years 1995 through 2004 (fiscal year 1995 was the earliest year for which financial information was still available). Only fires where the USFS was the recorded protection agency were used because of the difficulty of obtaining expenditures by all agencies involved in a wildfire (as discovered in the 1999 study). It was hoped that by making this restriction the USFS would have incurred the bulk of the expenditures on these fires, and the potential for underestimation due to not accounting for the expenditures of other agencies would be lessened. This seemed a reasonable assumption as the earlier study had indicated the USFS expended, on average, more than 90 % of the money on fires when they were listed as the protection agency in NIFMID (unpublished report on file at Rocky Mountain Research Station).

The name "Stratified Cost Index" stuck, though it was not a very descriptive name for the regression models that were developed.

Table 2.1 shows the list of variables used to develop the USFS expenditure models. In 2007, an article was published on the development of these models (Gebert et al. 2007). Also in 2007, RMRS conducted a study on the feasibility of developing similar expenditure models for the DOI agencies to respond to the congressional directive (Gebert 2007). The results of this study showed that comparable models could be developed for the DOI with similar predictive power to those produced for the USFS and development of models for the DOI was undertaken.

Though the language in the original Congressional report stated that the agencies should have a quantifiable performance measure, those involved in the development of the SCI suggested another approach due to the complexity of coming up with a single performance measure that takes into account the complexities of fire management. They suggested that rather than using the expenditure models to quantify the percentage of fires that exceeded the index, that those fires that ended up with expenditures more than one standard deviation above their expected expenditures be reviewed at the end of the fire season to determine the reason why these fires were so expensive and to learn from them. However, the agencies also needed a quantifiable performance measure, so the language in the Congressional report was simply amended to revise the "index" from

Table 2.1 Variables used in development of forest service expenditure regression equations; dependent variable = Ln(expenditures/acre)

Fire characteristics	Variable definition	Source
Size		
ln(total acres burned)	Natural log of total acres within the wildfire perimeter	NIFMID
Fire environment		
Aspect	Sine and cosine of aspect at point of origin in 45° increments	NIFMID
Slope	Slope percent at point of origin	NIFMID
Elevation	Elevation at point of origin	NIFMID
Fuel type	Dummy variables representing fuel type at point of origin. Grass = NFDRS fuel model A,L,S,C,T,N; Brush = NFDRS fuel model F,Q; Slash = NFDRS fuel model J,K,I; Timber = NFDRS fuel model H,R,E,P,U,G; brush4(reference category) = NFDRS fuel model B,O	NIFMID
Fire intensity level	Dummy variable for fire intensity level (FIL) category 1–6 (FIL 1 = reference category)	NIFMID
Energy release component	Energy release component calculated from ignition point using nearest weather station information (cumulative frequency)	Calculated
Palmer drought severity index	Average monthly PDSI for climate division containing fire ignition	NOAA
Values at Risk		
ln(distance to nearest town)	Natural log of distance from ignition to nearest census designated place[a]	Calculated
ln(total housing value 5)	Natural log of total housing value in 5 mile radius from point of origin (census data)	Calculated
ln(total housing value 20)	Natural log of total housing value in 20 mile radius from point of origin (census data)	Calculated
Reserved areas	Dummy variables indicating whether fire was in a wilderness area, inventoried roadless area, or other special designated area (reference category = not in reserved area)	Calculated
ln(distance to reserved area boundary)	If in a reserved area, natural log of distance to area boundary	Calculated
Detection time		
ln(detection delay)	Natural log of hours from ignition time to discovery time	Calculated
(ln(detection delay))^2	Square of ln(detection delay)	Calculated
Suppression strategy		
Initial suppression strategy	Dummy variables representing initial suppression strategy (confine, contain, control)—reference category = control	NIFMID
Resource availability		
ln(average deviation)	Natural log of the difference between the number of fires burning in the region during the period of the specified fire compared to the average in that region during the same time of year	Calculated
Region	Dummy variables for USFS region (reference category for western model = Region 1, for eastern model = Region 9	NIFMID

[a] All miles are air miles

Fig. 2.1 Percent of U.S. Forest Service large fires exceeding the SCI by 1 or 2 standard deviations, 2007–2012

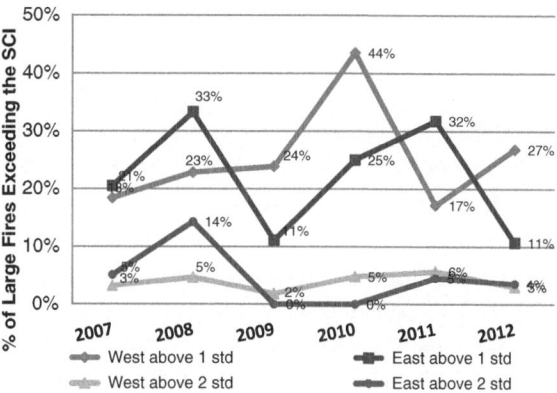

"expenditures per acre/energy release component" to "the percentage of fires, using a statistically representative sample, not contained in initial attack that exceed a 'stratified fire cost index.' This index would take into account known fire characteristics that affect expenditures; specifically, fire intensity, size, Forest Service region and proximity to communities, using historical expenditures per acre as a basis."

Hence, the use of the SCI as a performance measure for the Forest Service's expenditure containment efforts began, and since 2006, the SCI has been used to identify fires whose actual expenditures exceeded "benchmark" or estimated expenditures by more than one standard deviation (SD). The percentage of fires within one SD is meant to provide a rough measure of the effectiveness of the Forest Service's expenditure containment efforts. For use as a performance measure, the regression models used are essentially "stuck in time". That is, the regression models are based on fires from 1995 to 2004 and have not been updated. If the model were re-estimated each year with new data, the number of fires falling outside one or two standard deviations would not be expected to change. However, by staying with the original model, if expenditure containment measures are working, one would expect the distribution of large fires to shift to the left over time. Therefore, this leftward shift would cause fewer fires to show up as exceeding one standard deviation. In addition to the performance measure itself, fires that exceed one and two standard deviations are also likely candidates for further expenditure review after the fire season has ended.

To date, it is difficult to discern with certainty whether expenditure containment efforts are reducing the incidence of unusually high-expenditure fires. For large USFS fires from 2007 to 2012, there has been no clear trend in the percent of fires exceeding the SCI predictions by 1 or 2 standard deviations (Fig. 2.1). The SCI predictions are based only on comparisons of expenditures per acre, so it is possible that expenditure containment is affecting other fire outcomes, such as fire duration. However, given that the aggregate trend in expenditures has been increasing, there is no evidence to suggest that expenditure containment efforts have resulted in decreases in expenditures on large wildland fires. Trends for these outcomes are discussed in more detail in Chap. 3

2.3 Current Uses and Applications of Wildland Fire Expenditure Models

The suppression expenditure model described in the previous section is a useful tool for understanding how different environmental and socio-economic characteristics relate to wildland fire expenditures. To date, suppression expenditure models have supported a variety of planning and decision making tools. Expenditure models can be incorporated within a suite of fire and land management planning tools and can provide managers with valuable information about the consequences of fire management activities.

2.3.1 Fire Program Analysis

The Fire Program Analysis (FPA) is a federal inter-agency approach to planning and budgeting for a suite of fire management activities. The broad mission of FPA is to provide managers with a process and appropriate tools to make strategic budget and planning decisions for a wide range of fire management activities (Fire Program Analysis 2010). The potential expenditures on large fire suppression are only one aspect of FPA but have significant budgetary impacts. FPA incorporates suppression expenditure models as one method for examining the effectiveness of management activities. Well-specified expenditure models can provide detailed information about the effects of investments in fuel treatments and preparedness on large fire suppression expenditures.

Expenditure models are currently incorporated in FPA by applying a version of the SCI (Gebert et al. 2007) to large fires that arise under different management scenarios. The Large Fire Module of FPA is designed to examine how different management scenarios affect the likelihood, size, behavior (e.g., flame length), and costs of large fires (Fire Program Analysis 2012). Within this module, the suppression expenditure model can calculate differences in expenditures as the number of fires and their characteristics change under different scenarios. For example, the expenditure model would be used to estimate the change in expected suppression expenditures if a fuel treatment program results in a reduction in the expected number and intensity of large fires.

The use of expenditure models in FPA represents a shift toward using expenditure estimates as an indicator of cost effectiveness, rather than as a measure of performance itself. For example, one of the broad goals within FPA is to reduce the probability of occurrence of large and costly fires (Fire Program Analysis

2012). Reductions in large fire suppression expenditures are not a direct goal of FPA, but examining changes in suppression expenditures can highlight the budgetary consequences of reducing the incidence of large fires.

2.3.2   Decision Support: The Wildland Fire Decision Support System

Expenditure models are an important component of the Wildland Fire Decision Support System (WFDSS) currently used by federal wildland fire managers. WFDSS is an integrated suite of tools that allows fire managers to assess and manage risk, plan strategies with real-time incident information, and document decisions and management actions (Noonan-Wright et al. 2011; Pence and Zimmerman 2011). The primary purpose of WFDSS is to assist incident managers and other administrators in the development of risk-informed management strategies (Calkin et al. 2011b).

A part of the WFDSS suite of tools is a module to estimate the suppression expenditures for a given incident using an expenditure model based on the SCI. Models are separately developed for USFS fires that occur in the eastern and western regions, as well as for each DOI agency (National Park Service, Fish and Wildlife Service, Bureau of Indian Affairs, and Bureau of Land Management). The model parameters are updated each year by adding new data from the previous year to records dating back to fiscal year 2004, and include explanatory variables that are found to be significant for each agency and can be queried into WFDSS automatically using latitude and longitude points.[1]

The SCI implemented in WFDSS is a streamlined model that allows managers to quickly estimate a range of likely total suppression expenditures for a given incident. The original SCI model developed by Gebert et al. (2007) uses the fire ignition point to describe characteristics of each fire. In WFDSS, once an ignition point is available for an incident, the SCI module calculates the relevant characteristics to generate an expenditures prediction based on the likely size of the fire entered by the user. The output shows managers the median expenditures for their incident, as well as a range of potential expenditures.

The expenditures module in WFDSS serves multiple purposes for managers at various levels in fire management. First, expenditure estimates support budget planning by giving managers an initial estimate of how much similar fires have

---

[1] Some variables from the original SCI model would be difficult to include in WFDSS with acceptable accuracy, or in certain models are found to be insignificant predictors of expenditures. For example, fire intensity (flame length) and distance to the nearest town are omitted in all the WFDSS cost models, and region indicator variables are omitted in the National Park Service and Fish and Wildlife Service models. At a minimum, all of the models include fire size (entered by WFDSS users to generate model predictions) and fuel type at ignition (determined by ignition point latitude and longitude).

cost in the past. Second, expenditure estimates can serve as part of an early warning of incidents that are likely to be very costly. This may allow higher-level administrators to plan resource allocations based in part on potential expenditures. Additionally, estimates can be used by incident-level managers to identify when an ongoing fire is trending towards unexpectedly high expenditures. Managers can respond by making adjustments to ongoing strategies or improving the decision documentation of why current fire expenditures diverge from past fires (Pence and Zimmerman 2011).

2.4 Policy and Program Analysis: Assessing the Effects of Fuel Treatments and Forest Restoration on Suppression Expenditures

The link between wildfire suppression expenditures and hazardous fuel reduction is poorly understood. Though it is assumed that removing hazardous fuels from forests will reduce or remove the possibility of catastrophic wildland fires in areas that have been treated, even this basic premise is disputed by some, particularly in areas that historically have infrequent high–severity fires (Bessie and Johnson 1995; Schoennagel et al. 2004; Reinhardt et al. 2008). In fact Reinhardt et al. (2008) state "It is a natural mistake to assume that a successful fuel treatment program will result in reduced suppression expenditures. Suppression expenditures rarely depend directly on fuel conditions, but rather on fire location and on what resources are allocated to suppression. The only certain way to reduce suppression expenditures is to make a decision to spend less money suppressing fires". Also, Rideout and Ziesler (2008) suggest that wildland fire suppression and fuels treatments are not necessarily substitutes with clear tradeoffs. Investments in both treatments and suppression reduce fire damages and have complementary effects in reducing damages and expenditures.

In the real world, it is difficult to analyze the effect of fuel treatments on suppression expenditures because of the complexity of the factors influencing both wildfires and wildfire expenditures. However, modeling efforts, where some of these complexities can be controlled for, is one way to assess how fuel treatments may affect suppression expenditures. By linking models that predict fire extent and behavior, pre-and post-fuel treatment, with fire expenditure models such as the SCI, the possible effects of fuel treatments on suppression expenditures can be analyzed.

This type of modeling is currently underway, most notably in connection with the Collaborative Forest Landscape Restoration Program (CFLRP). An important objective of the CFLRP, created by the Omnibus Public Land Management Act of 2009, is to "facilitate the reduction of wildfire management costs, including through reestablishing natural fire regimes and reducing the risk of uncharacteristic wildfire." Given this emphasis on reducing fire management expenditures and the varying degrees of fire modeling and economic analysis capacity among CFLR teams, USFS economists and researchers came up with a standardized modeling approach and a spreadsheet tool [the Risk and Cost Analysis Tool (R-CAT)] to estimate expenditure savings from hazardous fuel management treatments funded by the program.

The modeling approach used in the R-CAT process combines the SCI model with the wildfire simulation model FSIM (Finney et al. 2011b) to predict pre- and post-treatment suppression expenditures. Estimated suppression expenditure savings are derived from the reduction in simulated final fire size due to treatment. These estimates are entered into the R-CAT spreadsheet, along with the costs of treatments and revenue generated, an estimate of the length of time that the treatments will be effective, and the timing of the treatments over the ten-year life of the project. The spreadsheet can then be used to calculate fire management savings, as opposed to suppression expenditure savings alone. Though the R-CAT process does not estimate changes in other fire management costs, such as changes in expenditures on small fires, reductions in Burned Area Emergency Response (BAER) costs, or potential reductions in per acre suppression expenditures due to being able to fight fire less aggressively in the treated areas, these potential savings can also be entered into the spreadsheet if there is sufficient justification for doing so.

The first pilot study of this effort was done for the Deschutes NF CFLR project and is described in more detail in Chap. 4 and Thompson et al. (2013d). Results showed a substantial reduction (around 35 %) in suppression expenditures. However, after accounting for the high costs of planned mechanical fuel treatments, overall fire management expenditures increased.

The results from the Deschutes pilot study highlight several important issues in terms of interpreting the results of any analysis related to hazardous fuel treatments and reductions in fire suppression expenditures. First, anticipated "savings" depend upon the metric being used. The success of hazardous fuel reduction in terms of saving money differs depending upon whether success is measured by suppression expenditure savings or fire program management savings. Fire program management savings are heavily influenced by the cost of treatments and the revenues generated from commercial activities, irrespective of suppression expenditure savings. Second, several other mechanisms by which suppression savings could be realized were not included explicitly in the Deschutes analysis including changes in fire intensity allowing for improved suppression efficacy, increased initial attack success leading to fewer large fires, and rehabilitation savings associated with less severe fires. For instance, results did show fewer large fires post-treatment, which was accounted for in the annualized results, but the effect on small fires per se was not analyzed. Finally, the results do not reflect "savings" in terms of reduced resource damage or enhanced ecological conditions, both important, if not primary, reasons for conducting hazardous fuel reduction.

Looking at the problem from a different perspective, Houtman et al. (2013) examined potential suppression expenditures savings and avoided timber losses associated with a "let it burn" policy rather than an active suppression policy. In that study, they integrated a wildfire spread model (Finney 2004) with a forest vegetation growth model to simulate wildfire spread coupled with changes in vegetation conditions. Using a study area on the Deschutes NF, several hundred different sample pathways, in terms of fires and growth events, were simulated to assess the effect on suppression expenditures in the future. Future suppression expenditures for large fires (greater than 121 hectares) were estimated using the SCI model. Smaller fires were estimated using average expenditures for the Deschutes NF. Though the results varied, some of the potential future scenarios did demonstrate a higher present value of the landscape resulting from the let-burn management choice, particularly if not accounting for the value of the lost timber. Not surprisingly, the greatest benefits in terms of suppression expenditure savings occurred when the current fire was large and occurred early in the time horizon.

2.5 Explaining Fire Management Decisions with Expenditure Models as a Research Tool

Suppression expenditure models have made significant advances in describing relationships between fire expenditures and landscape characteristics, geography, and socio-economic factors. Yet at a basic level, expenditures reflect management decisions to deploy suppression resources when responding to a dynamic spatio-temporal fire environment. Expenditure models are increasingly able to give researchers and managers a window into these decisions and allow for tests of hypotheses related to economic decision making and resource allocation in a wildland fire context.

2.5.1 Management Decision Making and Expenditures

Because suppression expenditures reflect management decisions at multiple levels to assign resources to a fire incident, suppression expenditure models may yield insights into the underlying mechanisms that drive resource allocation decisions. A better understanding of how management decisions are made can suggest strategies and tools to improve decisions and the cost-effectiveness of fire management.

A key area of management decision making that may determine expenditures is the choice of overall suppression strategy. Strategies may vary broadly from aggressive direct suppression efforts meant to contain a fire quickly and limit damage, to monitoring efforts that largely allow the fire to take its natural course when few highly valued resources are at risk. These types of strategies will call for different time-paths of resource allocation as managers weigh tradeoffs over fire size, duration, intensity, ecological impact, and damage to valued assets.

Gebert and Black (2012) examined the suppression strategies ranging from direct (full) suppression to strategies that primarily seek to allow the fire to achieve resource benefits. As expected, fires using the most aggressive strategies tended to result in shorter duration fires with smaller burned areas, but with higher expenditures per acre and per day (due to more intensive suppression effort over a shorter time and smaller area). However, total suppression expenditures were about the same for fires using the least aggressive suppression strategy compared with those using full suppression strategies; the longer duration of less aggressive suppression efforts offset the per acre and daily expenditures savings.

The relationships between strategy choice and expenditures (and other fire outcomes) can provide some basic guidance to managers on the consequences of different strategies. However, they also highlight that expenditures alone cannot capture all of the factors that managers consider when making strategic decisions. Suppression expenditures can be incorporated into models of optimal suppression strategies to provide a link between expenditures, fire growth, and damage or benefits caused by the fire (Gorte and Gorte 1979; Donovan and Rideout 2003). Petrovic and Carlson (2012) showed that incorporating the effects of different suppression strategies on expenditures (Gebert and Black 2012) can help identify priority areas for intensive suppression and help to efficiently allocate suppression resources between fires.

Expenditure models can also provide evidence of whether the tools provided to managers are effective. The intent of decision support systems such as WFDSS is to improve overall efficiency and risk management during an incident (Calkin et al. 2011b), and not necessarily constrain expenditures. Hesseln et al. (2010) suggest that the use of geo-spatial tools, such as digital fire progression maps and geo-spatial fire analyst units, during an incident may improve efficiency but not necessarily reduce expenditures. They find that expenditures were not statistically different for incidents where geo-spatial tools were used but tended to be closer to an efficient cost-minimizing frontier when compared with other incidents. Geo-spatial tools may improve information about fire behavior and landscape characteristics but can be costly investments. Results suggest that suppression resources are being more efficiently allocated to achieve better fire outcomes at a given level of expenditures.

2.5.2 External Factors and the Socio-Political Environment

Within fire management organizations, incentive structures, policies and regulations, and socio-political pressure can affect how decisions are made during an incident. Factors external to the conditions and environment for a given incident may play a large role in how resources are allocated within and between incidents. Even after accounting for many of the biophysical factors and landscape characteristics that are related to fire expenditures, much variation in expenditures remains unexplained. This suggests that human factors including the characteristics of the local land managers and assigned incident management team as well as social and political influences may play a substantial role in determining fire expenditures.

Because the allocation of suppression resources to a fire incident involves a series of decisions by individual managers, examining the incentives that drive manager decisions may help explain variations in expenditures. Donovan and Brown (2005) suggest that manager incentives are a contributing factor resulting in the build-up of hazardous fuels that cause larger, more dangerous, and more costly fires. This line of reasoning suggests that a budgeting process whereby expenditures are paid largely from national accounts rather than from local accounts may encourage overuse of suppression resources on a given fire and over the course of a fire season. This limits the amount of fire on the landscape that can naturally treat and reduce hazardous fuels, which leads to larger and more expensive fires in the future. The authors identified two factors that create an incentive for overuse of suppression resources. First, budgets for suppression are paid out of national funds, rather than from budgets at the local level; a manager in a local unit does not bear any of the opportunity cost for using additional suppression resources because they are not paid out of the local fund. Second, management decisions tend to discount beneficial effects of fire for treating the build-up of fuels. That is, managers may only consider the potential damages of additional fire on the landscape and not the potential future benefits of a less aggressive suppression response that allows for more extensive burning.

The disincentives faced by managers to contain expenditures have been recognized as a potential barrier to more efficient risk management of wildland fire (Calkin et al. 2011d). However, the focus on suppression budgeting in Donovan and Brown (2005) may understate the problem. Risk management for wildland fire includes a broad array of land management activities, and incentive-compatible suppression budgeting may not address incentives for other proactive measures to reduce risks from fire (Thompson et al. 2013a).

Thompson et al. (2013a) suggest that risk management for wildland fires could benefit from the application of actuarial principles to fire program funding. An actuarial approach could encourage both suppression expenditure containment and efficient management of risk. Suppression expenditure models may be an important component of this process; Thompson et al. (2013a) use a version of the expenditure model developed by Gebert et al. (2007) to estimate suppression expenditures for simulated fires on national forest units in two regions. The distribution of expenditures on simulated fires shows that significant differences in expenditures can be expected between and within regions. These results indicate that suppression expenditure modeling can provide valuable information about the expected benefits of fire program investments.

Expenditure models may help reveal how the incentives faced by managers can be altered by social and political pressures from outside the fire management organization. Fire managers themselves recognize that such pressures exist and can affect suppression expenditures (Canton-Thompson et al. 2008). A common example of this is the relationship between housing and other private property and suppression expenditures. Development in Wildland-Urban Interface (WUI) areas has resulted in more people and homes within or closer to fire-prone areas, particularly in the western United States. This trend has been cited as a primary factor that increases suppression expenditures (OIG 2004). The U.S. Forest Service and other federal fire management agencies do not have a specific mandate to protect private property; however, social and political pressure from local residents and their representatives may encourage managers to increase suppression efforts.

The proximity of development and private property tends to be associated with higher suppression expenditures. The regression-based models of suppression expenditures show that total and per-acre expenditures tend to be higher when more housing is closer to fire incidents (Gebert 2007; Yoder and Gebert 2012) or when private land is a greater share of the burned area (Liang et al. 2008). But these studies provide only indirect evidence that manager incentives are being altered through social pressure to increase suppression response. The results may represent an effort by managers to increase investments in suppression when highly valued resources are at risk (and which are likely correlated with development in WUI areas).

A more direct investigation of the effect of social and political pressure was carried out by Donovan et al. (2011). Using regression-based expenditure models, the authors associated the volume of newspaper coverage of fire incidents and the seniority of Congressional representatives in the district where incidents occurred to suppression expenditures. They found that more newspaper coverage and more senior Congressional representatives were associated with higher expenditures, even after controlling for other fire characteristics that typically drive expenditures.

To the extent that social and political pressure plays a role in determining suppression expenditures, it likely operates through altering manager responses to risk and the tradeoffs managers make over different fire outcomes (including expenditures). For example, Calkin et al. (2013) found that when choosing among suppression strategies, managers appear to place little weight on suppression expenditures when homes are at risk and may actually seek out more expensive strategies. Further, the authors found that there is a divergence between choices managers would likely make under realistic conditions, and the choices they would prefer to make if social and political pressures were absent. A follow-up study found that manager decisions may be tied in large part to risk preferences (Wibbenmeyer et al. 2013).

2.6 Conclusion

The Stratified Cost Index (SCI) and subsequent expenditure models have been developed largely in response to large and increasing federal expenditures for wildland fire management. As fire management expenditures have increased over the past decade, so too have calls to better understand the factors that may explain expenditures. This understanding can help determine which factors may be under management control and how agencies involved in wildland fire management can more cost-effectively manage fires.

Expenditure models will need to continue to provide improved links between expenditures, management strategies, and fire outcomes. Indeed, much of the recent work on expenditure modeling has sought to better link expenditures (and other fire outcomes) to the decision making processes that commit suppression resources to fire management efforts (Gebert and Black 2012; Donovan et al. 2011). Expenditure models are ultimately an attempt to evaluate management strategies in terms of expected net value change, including suppression expenditures, resource damages, and risks to human life and safety. Donovan and Rideout (2003) identify that net value change for a fire is dependent on pre-determined levels of pre-suppression and fuels management investments. Thus, a goal of expenditure modeling is to better understand how suppression efforts change large fire spread and potential outcomes to highly valued resources, the interactions of suppression spending on pre-suppression and fuel treatment levels, and how firefighter and public safety is affected by alternative strategies.

The need for accurate and reliable expenditure models is only likely to increase in the future. (Chapter 3 in this volume explores past performance issues with expenditure models, and Chap. 4 describes advances in model development to improve accuracy.) Increased development, particularly in the western United States, will result in larger populations in closer proximity to forests (USDA Forest Service 2012). Climate change is likely to increase frequency and severity of fires in these same geographic areas (Westerling et al. 2006), which could result in a large increase in the exposure of human populations to risks of wildland fire.

Increases in exposure to wildland fire could precipitate a concomitant increase in expenditures. Given that fire management budgets already account for large portions of budgets for agencies involved in wildland fire management, expenditure models may be able to play an important role in improving cost effectiveness in the future. However, it is also possible that development and climate changes will alter the expenditure relationships that have been discovered since the original SCI was first developed. That is, what we currently know about how fires are managed and their costs may change in response to the altered relationships between human populations and fire-prone landscapes.

Chapter 3
Regional and Temporal Trends in Wildfire Suppression Expenditures

Abstract This chapter explores regional and temporal patterns of wildfire suppression expenditures. A key question addressed is whether expenditure differences across time or between regions are explained by differences in fire characteristics, differences in management responses to fire, or unobserved factors. Three analyses are conducted: A time-series analysis of aggregate expenditures at the regional level to discern year-over-year trends due to random processes; an analysis of trends in cost-performance metrics based on the Stratified Cost Index model; and, a micro-level regression model that examines the determinants of suppression expenditures over time and between regions. In the latter case, introducing year- and region-specific interactions with fire characteristics in expenditure regressions can indicate whether observations of higher (or lower) expenditures in certain years and regions can be attributed to variation in fire characteristics or to unobserved differences between years and regions. Taken together, these analyses highlight the complex relationships between expenditures, fire characteristics, climate and weather, and human factors in determining suppression expenditures.

Keywords Expenditure trends · Expenditure performance measures · Stratified cost index · Regression decomposition · Regional trends · Augmented Dickey-Fuller · Trend · Drift · Random walk

3.1 Introduction

This chapter explores regional and temporal patterns of suppression expenditures. Despite the constant presence of fire on the landscape, fire seasons can exhibit enormous variations in the extent and severity of fires, the effects of wildfires on humans, and the costs of managing fires. Further, regional variations in climate and weather ensure that management of wildfires can vary between regions and from one year to the next.

M. S. Hand et al., *Economics of Wildfire Management*, SpringerBriefs in Fire,
DOI: 10.1007/978-1-4939-0578-2_3, © The Author(s) 2014

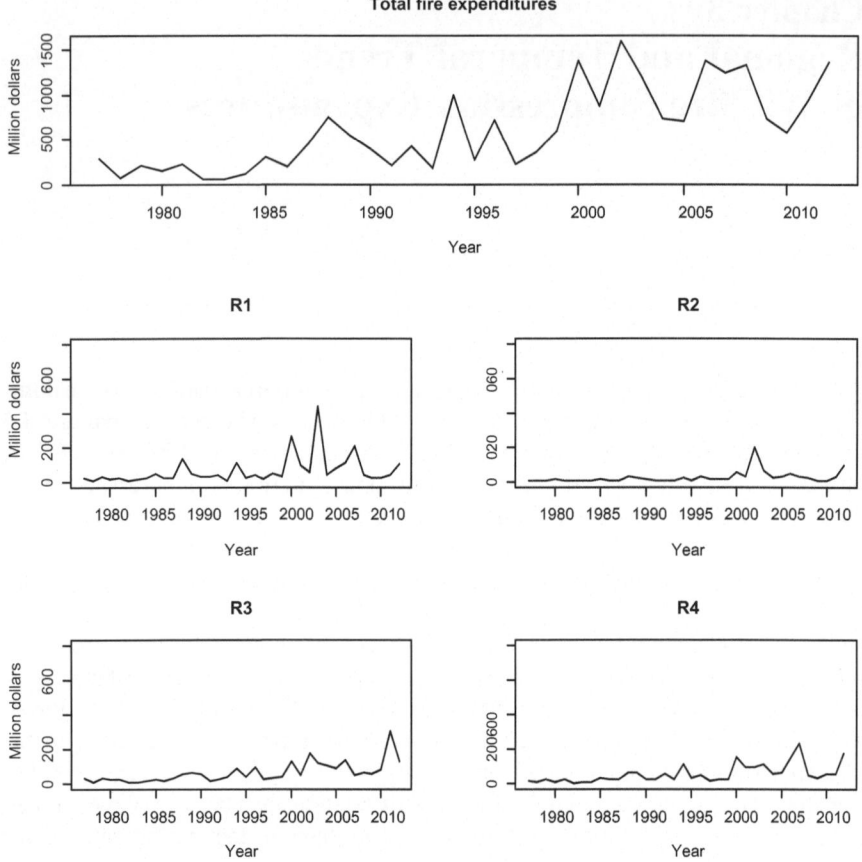

Fig. 3.1 Aggregate fire suppression expenditures from 1977 to 2012

Examining U.S. Forest Service expenditures suggests that there are significant variations in expenditures between regions and across time (Fig. 3.1). Total expenditures are higher over the most recent decade than in the previous decade, although there doesn't appear to be a consistent year-over-year trend towards increasing expenditures. At a regional level, Region 5 (California) tends to have the largest expenditures in any given year. However, regional "hot spots" in a particular year may push other regions higher if fire activity is particularly heavy in a region. For example, Region 6 (Northwest) in 2002, Region 1 (Northern Region) in 2000 and 2003, and Region 3 (Southwest) in 2011 all showed spikes in expenditures corresponding to heavy fire activity.

To provide a more detailed analysis of regional and temporal trends, aggregate expenditures are examined using time-series statistical techniques and the Stratified Cost Index (SCI) model. Of particular interest is whether differences in average expenditures across regions and years can be explained by differences in

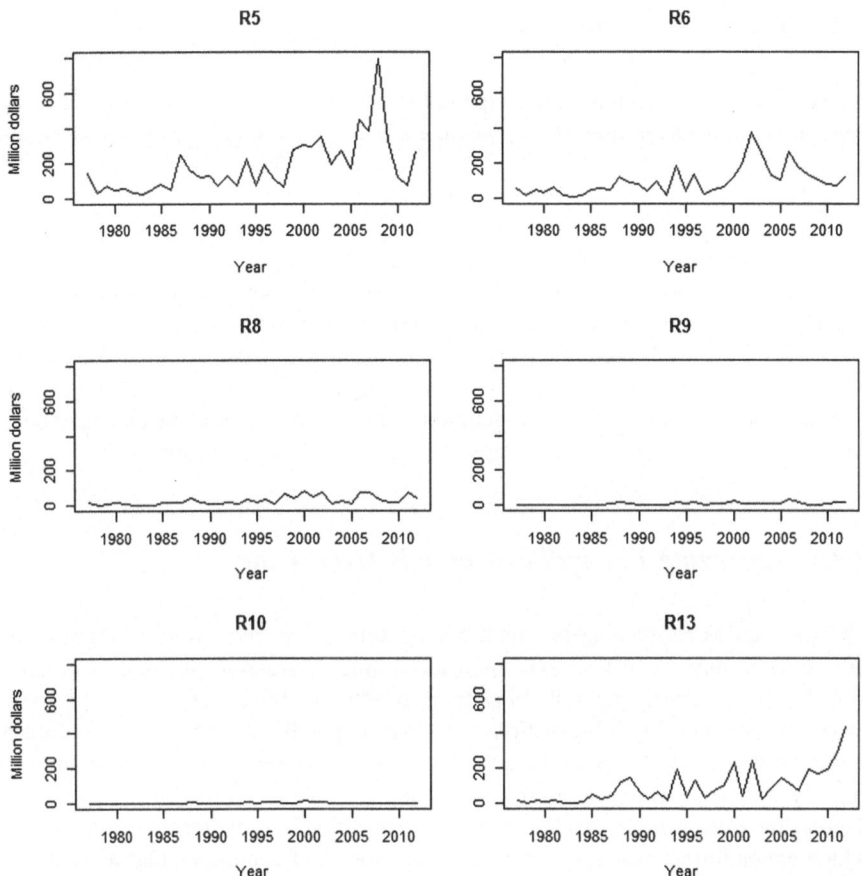

Fig. 3.1 continued

fire characteristics (e.g., fire conditions becoming more severe over time), or by differences in the relationships between fire characteristics and expenditures (e.g., a more costly response to changes in fire conditions). The first explanation (conditions becoming more severe) is biophysical in nature, whether the increase is due to increasing fuel loads from past suppression efforts or from the effects of climate change. As such, it would require a different set of management responses geared more toward manipulation of or adaptation to the biophysical environment. The second is more "human" centered (responses to increased settlement near fire-prone landscapes, social and political pressures, etc.); leading to management responses geared more toward changing human behavior, organizational incentives, and institutional controls.

3.2 Methods and Data

To examine trends in suppression expenditures over time and between regions, two general techniques are used. First, we take a broad look at aggregate expenditures using detailed time-series analysis at the regional level. Time-series techniques can help discern year-over-year trends in expenditures from changes due to random processes and are useful for describing the behavior of expenditure observations (if not the structural determinants of observed trends).

The second approach uses micro-level data to examine how fire characteristics that are related to expenditures may explain trends over time and between regions. We use the results from SCI-model predictions of per-acre fire expenditures to determine whether expenditures are increasing or decreasing relative to historical averages. We also develop a version of the SCI regression model to explore how fire characteristics can explain differences in expenditures over time and between regions.

3.2.1 Aggregate Expenditure Trends Over Time

The time-series analysis tests whether expenditures are increasing or decreasing over time in any (or all) regions after accounting for random factors. More specifically, we examine the following three questions about trends in suppression expenditures: (1) Do observations of past expenditures help predict future expenditures? (2) Do expenditures tend to increase or decrease by a fixed amount each year? (3) Are year-over-year changes in expenditures increasing or decreasing over time? Answering the first question can indicate the degree to which expenditures change according to decisions and conditions that determined expenditures in previous years. The second question can show whether expenditures are moving in a certain direction regardless of decisions and conditions in previous years. Finally, the third question is useful for understanding whether expenditures follow a trend that is increasing or decreasing over time.

A series of three augmented Dickey-Fuller tests are used to answer the questions posed above: (a) a random walk; (b) a random walk with a drift; (c) a random walk with a drift and deterministic trend (Pfaff 2008). The tests are designed to indicate whether expenditures over time behave as a stationary process. A stationary time series implies that expenditures over time are random around a fixed mean and with a fixed standard deviation, whereas a non-stationary series exhibits expenditures with a non-constant mean or standard deviation (or both).

A random walk process of regional expenditures can determine whether past realizations of expenditures predict future expenditures:

$$\Delta C_{kt} = \gamma C_{k,t-1} + \sum_{i=1}^{p-1} \delta_{ki} \Delta C_{k,t-i} + e_{kt}, \tag{3.1}$$

where ΔC_{kt} represents the first difference (i.e., $C_{kt} - C_{k,t-1}$) of aggregate expenditures ($) of region k in year t, γ and δ_{ki} are parameters to be estimated, and p is the order of autoregression. The γ parameter can be interpreted as the effect of last year's aggregate expenditures on the changes in expenditures observed this year, and the δ_{ki} parameters indicate the effect of previous years' changes in expenditures on this year's observed change in expenditures. The number of previous years to include in the regression (i.e., the order of autoregression, p) was selected using Bayesian information criterion to find the model that best fits the data.[1]

To test whether expenditures increase or decrease by a fixed amount each year, a constant term β_0 is added to Eq. (3.1), resulting in:

$$\Delta C_{kt} = \beta_0 + \gamma C_{t-1} + \sum_{i=1}^{p-1} \delta_i \Delta C_{k,t-i} + e_{kt}. \tag{3.2}$$

Including a deterministic time trend requires adding a parameter (β_1) that describes the increase or decrease in expenditures that occurs over time (t):

$$\Delta C_{kt} = \beta_0 + \beta_1 t + \gamma C_{k,t-1} + \sum_{i=1}^{p-1} \delta_{ki} \Delta C_{k,t-i} + e_{kt}. \tag{3.3}$$

All three models were estimated with Feasible General Least Squares procedure (Greene 2003, p.209) using regional and total USFS suppression expenditures from 1977 to 2012, adjusted for inflation.

The null hypothesis of the stationarity test is the same for each equation: $\gamma = 0$. Rejecting the null hypothesis indicates that the series is stationary and observations of past expenditures help explain future changes in expenditures. Not rejecting the null hypothesis (i.e., the time series is non-stationary and has a unit root) indicates that knowledge of past expenditures does not help predict future changes in expenditures. If the null hypothesis is not rejected, tests for the significance of the drift and/or deterministic trend models can be performed using Eqs. (3.2) and (3.3).[2]

3.2.2 Regional Spatial Relationships in Expenditures Over Time

To investigate the potential presence of spatial relationships among regional expenditures over time, we examine the relationship between aggregate expenditures of border-sharing (neighboring) regions. One way to study this relationship

---

[1] An autoregressive order for each region between 2 and 5 best fits the data according to the BIC. Results available from the authors upon request.

[2] These tests are a direction for future research. Details of the tests for drift and deterministic trends can be found in many time-series textbooks, for example Enders (2008). For the purposes of testing for stationarity, the estimated parameters of the trend, drift, and past changes do not have an interpretable meaning like they do in conventional time series models, for example autoregressive moving average models. Including past changes in Eqs. (3.1), (3.2), and (3.3) ensures the model residuals are white noise with no autocorrelation.

quantitatively is to construct a spatial neighborhood structure and estimate a space–time model of the first differences with a three-stage iterative approach (Pfeifer and Deutsch 1980). However, this kind of model usually involves the estimation of a large quantity of parameters due to the neighborhood structure as well as the orders of autoregression and moving average terms. With only 35 observations in each series of differenced expenditures, excessive loss of degrees of freedom is inevitable, even if only first-order spatial neighbors were considered.

The approach taken here is to conduct an Engle-Granger test for co-integration (Engle and Granger 1987). This test can determine whether the expenditures in two neighboring regions tend to move together in the long term (i.e., whether there is a long-run relationship between expenditures in the two neighboring regions[3]). Co-integration tests have been used extensively to test how closely the prices in different markets move together and provide evidence for spatial market structure (Goodwin and Piggott 2001; Goodwin and Schroeder 1991). In the present case, co-integration can indicate whether spatial relationships have an influence on aggregate suppression expenditures at the regional level.

The test is comprised of two steps: 1. Estimate Ordinary Least Squares (OLS) for the following regression: $C_{it} = \theta C_{jt} + e_t$, 2. Perform stationarity test on the residuals: $\hat{e}_t = C_{it} - \hat{\theta} C_{jt}$, with an augmented Dickey Fuller test that contains a drift and trend:

$$\Delta \hat{e}_t = \rho_0 + \rho_1 t + \alpha \hat{e}_{t-1} + \sum_{i=1}^{k-1} \mu_i \Delta \hat{e}_{t-i} + u_t \tag{3.4}$$

Here, C_{it} is the aggregate expenditure of region i in year t and $\theta, \rho_0, \rho_1, \alpha,$ and μ_i are parameters to be estimated. If $\alpha \neq 0$ ($\hat{e}_t$ is stationary), then expenditures in the two regions are co-integrated, i.e., they move together in the long run.

3.2.3 Regional and Temporal Relationships in the SCI Model

The SCI model can be used to examine trends in individual fire expenditures over time and between regions. We examine predicted fire expenditures and SCI model parameter estimates to answer two broad questions:

- Has expenditure containment performance improved over time, and are there significant differences in performance between regions?

---

[3] Region 10 and Region 13 are omitted from the spatial co-integration tests. Region 10 is comprised only of Alaska, and thus doesn't share any contiguous borders with other regions. Region 13 is an identifier for suppression expenses that originate from the Washington, DC headquarters or are otherwise not assigned to a specific geographic region. Expenditures associated with Region 13 tend to be high, because national resources, such as helicopters and air tankers, as well as national contracting expenses, such as catering services, are charged to Region 13.

- Are differences in expenditures due to different characteristics of fires or to different relationships between explanatory factors and expenditures?

To answer the first question, annual expenditure performance metrics for large fires managed by the USFS are compared between regions and over time. The original SCI model was developed in part to provide a broad performance measure for USFS expenditure containment efforts (see Chap. 2). That is, the SCI can be used to measure which fires in a given season were more or less expensive than average fires with similar characteristics and whether expenditure containment efforts are effective over time. The SCI model described in Gebert et al. (2007) closely matches the model used by the USFS for assessment of expenditure performance, where actual expenditures per hectare (EPH) are compared to model-estimated expected EPH.

The expenditure assessment model measures the outcome of each large fire (greater than 121 hectares) as the total agency expenditures per hectare. For each fire, the SCI regression model parameters are used to calculate the expected EPH for any fire based on the fire characteristics in the model, and to compare expected EPH to actual EPH. Using the standard deviation in EPH from the estimation sample, fires are identified where actual EPH exceeds expected EPH by one or two standard deviations. The frequency of fires exceeding one or two standard deviations in EPH is summarized by region for fiscal years 2006–2012.

To answer the second question, a decomposition of SCI-type regression parameters is used to provide insight into the sources of previously unexplained expenditure differences between regions and across time. Previous studies have highlighted categorical differences in expenditures between regions and over time. For example, suppression expenditures per hectare are higher on average for fires in California and the Northwest Region (Washington and Oregon) than in other regions (Gebert et al. 2007). Donovan et al. (2011) found that expenditures vary between regions, although the differences for California and the Northwest are no longer significant after controlling for media coverage and Congressional seniority. Yoder and Gebert (2012) found that expenditures are higher in certain years but do not appear to follow a consistent time trend.

Because previous studies controlled for other factors thought to drive expenditures, regional and temporal expenditure differences may be accounted for by unobserved variables or structural differences in how fires are managed. A simple pooled regression of USFS and DOI large fires illustrates the differences in expenditures per hectare by landscape characteristics, geographic region, administrative agency, and year. Interaction terms for key categorical variables are introduced to examine the degree to which categorical differences are due to structural differences in how expenditures are determined. These interactions can indicate where structural heterogeneity is an important determinant of expenditures and where remaining unobserved factors account for expenditure differences.

The regression model is based on the SCI model in Gebert et al. (2007), and related studies using similar techniques (e.g., Liang et al. 2008). In addition to landscape characteristics, the model controls for geographic regions, year, and administrative agencies with primary management responsibilities. The basic model is expressed as:

$$EPH_i = \beta X'_i + \gamma DOI_i + \delta G'_i + \alpha FY'_i + \varepsilon_i, \tag{3.5}$$

where EPH_i is an observation of the natural log of total federal expenditures per hectare for fire i, X_i is a vector of landscape characteristics (including a constant) calculated at the ignition point of each fire, DOI_i is a categorical variable indicating fires where a DOI agency had primary management responsibilities (instead of USFS), G_i is a vector of geographic area categorical variables, and FY_i is a vector of fiscal year categorical variables. The ε term is a random disturbance that is assumed to be distributed with a mean of zero and constant variance.[4]

The geographic area (G) and year (FY) categorical variables are of primary interest for examining structural differences in the expenditure regression. For a categorical variable with a statistically significant regression coefficient in the basic model, interactions with the other model variables can be used to test the hypothesis that category-specific regression coefficients are a significant determinant of expenditure differences. The interaction model for a generic categorical variable is:

$$EPH_i = \beta X'_i + \varphi\left(Z'_i \times C_i\right) + \varepsilon_i, \tag{3.6}$$

where C_i is a generic categorical variable, X_i includes all landscape characteristics and categorical variables (including C_i), and Z_i is the subset of variables from X_i that are interacted with C_i.

Individual interaction coefficients in φ can be tested for significance using conventional t-tests. These tests can be interpreted as showing whether the expenditure response to a particular fire characteristic is different for the category of interest as compared with the rest of the population. However, it may be that the regression model as a whole is significantly different for the category of interest. Testing the hypothesis that all of the interaction terms are jointly significantly different from zero is carried out with a likelihood ratio statistic (Greene 2003):

$$LR = -2 \times ln\left(\frac{L_r}{L_u}\right), \tag{3.7}$$

where L_u is the value of estimated likelihood function for the full (unrestricted) model in Eq. (3.6), and L_r is the value of the estimated likelihood function for the basic (restricted) model in Eq. (3.5) (where all of the interaction terms are

---

[4] The assumption of spherical disturbances appears to be a reasonable one for this application; Gebert et al. (2007) and Liang et al. (2008) did not find evidence of heteroskedasticity or (in the latter study) spatial auto regressive errors. In the present study, controlling for a general non-spherical disturbance process (using White-Huber sandwich estimator) resulted in very little change in standard errors and no change in the interpretation of results. Results available from the authors upon request.

restricted to equal zero). The test statistic is Chi squared distributed with degrees of freedom equal to the number of restrictions (i.e., the number of parameters in φ in Eq. (3.6)).

To aid in the analysis of the categorical variables, and to avoid specifying an unwieldy number of interaction terms, the geographic areas and years are grouped into low-, moderate-, or high-cost categories based on the initial regression model from Eq. (3.5). Consistent with previous studies, initial regressions identified California and the Northwest as the high-cost regions. Moderate-cost regions include the Southwest and Great Basin regions; the Northern Region, Rocky Mountain, Eastern, and Southern regions make up the low-cost regions.[5] High-cost years include 2004, 2010, 2011, and 2012, moderate-cost years include 2005 and 2008, and low-cost years include 2006, 2007, and 2009. In the interaction specification (Eq. 3.6), the high-cost categories for geographic areas and years are interacted with the other variables in the regression.

The regression model is estimated using a dataset of large fires managed by either the USFS or a DOI agency. Expenditure data (in constant 2012 dollars) are drawn from USFS and DOI agency financial databases and matched with geographic data from the ignition point of each fire using Geographic Information System (GIS) layers maintained by the Wildland Fire Decision Support System (WFDSS). The final data set is a combined set of observations that are used to generate the expenditure models in the WFDSS SCI module in use for the FY 2013 fire season.[6] Table 3.1 describes the data used for estimation.

3.3 Results

3.3.1 Aggregate Expenditure Trends Over Time

As shown in Table 3.2, the hypothesis that expenditures are non-stationary ($\gamma = 0$) cannot be rejected for at least two of the three models in all cases. We also found that for all models where γ is not significantly different from zero, the drift and deterministic trend were also not significant at conventional confidence levels. Thus, annual aggregate expenditures appear to be non-stationary and best described as random walks, which is consistent with the observed upward change over time. This implies that at the aggregate level, (1) the mean and/or variance of fire suppression expenditures changed over time from 1977 to 2012, and, (2)

---

[5] For the purposes of this analysis observations in Alaska are omitted. Relatively few Alaska fires are managed by the USFS, which confounds results with respect to the DOI categorical variable. Also, it appears that an expenditures model for Alaska may vary significantly from a model for the contiguous 48 states.

[6] The WFDSS SCI model includes geographic area categorical variables, but does not control for differences between years.

Table 3.1 Descriptions of regression variables

Variable	Description	Mean	Std. Dev.
LNEPH (dep. var.)	Natural log of total federal expenditures per hectare burned, adjusted for inflation (2012 $)	5.5	1.8
LNHECTARE	Natural log of total burned hectares	5.81	1.62
ERC	Relative energy release component (0–100 scale)	81.1	19.7
GRASS	Binary indicator of grass fuels at ignition	0.46	0.498
BRUSH_FMD	Binary indicator of brush (chaparral) or fuel-model D (Southern rough) fuels at ignition	0.269	0.444
BRUSH4	Binary indicator of brush-4 (dense brush) fuels at ignition	0.034	0.18
TIMBER	Binary indicator of timber fuels at ignition	0.237	0.425
LNELEV	Natural log of elevation at ignition	6.76	0.824
LNTOT20	Natural log of total population within 20-mile radius of ignition	7.31	3.1
WILD	Binary indicator of whether ignition point is within a designated wilderness area	0.064	0.244
LNWILD_DIST	Natural log of distance from ignition point to nearest wilderness area boundary (if WILD = 1)	0.054	0.252
LNTOWN_DIST	Natural log of distance to the nearest town with population of at least 50,000	2.44	0.76
DOI	Binary indicator of whether fire was managed by a DOI agency	0.78	0.414
LOW_COST_GEOG	Low cost regions (binary): Northern, Rocky Mountain, Eastern, Southern	0.397	0.489
MID_COST_GEOG	Moderate cost regions (binary): Southwest, Great Basin	0.419	0.493
HIGH_COST_GEOG	High cost regions (binary): California, Northwest	0.184	0.388
LOW_COST_YEAR	Low cost fiscal year (binary): 2009	0.415	0.493
MID_COST_YEAR	Moderate cost fiscal years (binary): 2005, 2008	0.226	0.418
HIGH_COST_YEAR	High cost fiscal years (binary): 2004, 2010–2012	0.359	0.48

expenditures exhibited an unpredictable path, so knowledge of past expenditures would not help predict future movements.[7]

There are 15 pairs of regions that share borders, and eight pairs are co-integrated at the 95 % confidence level: Region 1 and Region 2, Region 1 and Region 4, Region 1 and Region 6, Region 3 and Region 5, Region 3 and Region 8, Region 4 and Region 5, Region 4 and Region 6, and Region 5 and Region 6 (Table 3.3). This result implies that the movement of expenditures in Region 2 was relatively independent of the surrounding regions while expenditures in Region 1 seemed to

---

[7] Dickey-Fuller tests may suffer from a lack of statistical power in some cases, which may account for the small number of rejections (DeJong et al. 1992). Results confirm that differencing the expenditures time series (i.e., so the dependent variable is the change in expenditures) is necessary for spatial and non-spatial time-series modeling to avoid spurious regression. Stationary tests performed on the first differences of all series indicated that all expenditures series were integrated of order 1 (i.e., were stationary).

Table 3.2 Test statistics for Augmented Dickey-Fuller stationary tests of regional and total fire suppression expenditures (C) from 1977 to 2012

Random walk (Eq. 3.1)	τ^a	p		
R1	−1.1273	3		
R2	−1.709	2		
R3	0.054	2		
R4	−1.0943	2		
R5	0.8494	5		
R6	−1.1352	2		
R8	−0.3155	4		
R9	0.031	5		
R10	−1.7865	2		
R13	1.2144	2		
Total	−0.2415	2		
Drift (Eq. 3.2)	τ	$\Phi1^b$	p	
R1	−2.094	2.2094	3	
R2	−2.6988	3.748	2	
R3	−1.237	1.1736	2	
R4	−2.7243	3.8752	2	
R5	−2.3055	2.6954	2	
R6	−2.2763	2.6067	2	
R8	−1.6363	1.4722	4	
R9	−2.0651	2.3364	4	
R10	−2.5574	3.2704	2	
R13	0.1715	0.8943	2	
Total	−1.6639	1.6546	2	
Drift and trend (Eq. 3.3)	τ	$\Phi2^c$	$\Phi3^d$	p
R1	−3.8786*	5.0393	7.5436*	2
R2	−3.2315	3.5992	5.2856	2
R3	−3.1771	4.0201	5.5044	2
R4	−4.4283*	6.7462*	9.9022*	2
R5	−4.9254*	8.1874*	12.2246*	3
R6	−2.7409	2.5412	3.7951	2
R8	−5.2656*	9.329*	13.8771*	3
R9	−6.6854*	15.1129*	22.4376*	3
R10	−2.4429	2.1286	3.1928	2
R13	−1.5476	2.0539	2.1010	2
Total	−4.3216*	6.5041*	9.3609*	3

Notes p indicates autoregressive order selected using minimum BIC

* indicates significance at 95 % confidence level

[a] Hypothesis: $\gamma = 0$. Critical values at 95 % confidence level for τ in the three models are −1.95, −2.93, −3.50, respectively (McLeod et al. 2011)

[b] Hypothesis: $(\beta_0, \gamma) = (0, 0)$. Critical value at 95 % confidence level is $\Phi1^* = 4.86$

[c] Hypothesis: $(\beta_0, \beta_1, \gamma) = (0, 0, 0)$. Critical value at 95 % confidence level is $\Phi2^* = 5.13$

[d] Hypothesis: $(\beta_1, \gamma|\beta_0) = (0, 0|\beta_0)$. Critical value at 95 % confidence level is $\Phi3^* = 6.73$

For concision, β_0, β_1, and δ_i are not shown in the table

Table 3.3 Test statistics for co-integration tests (Eq. 3.4) of fire suppression expenditures of border-sharing neighbors

	R1	R2	R3	R4	R5	R6	R8	R9
R1		−5.928		−3.5141		−6.1839		−3.3838
R2			−3.3859	−3.4905			−3.228	−3.3976
R3				−1.3359	−5.6297		−6.372	
R4					−3.7034	−4.0353		
R5						−4.9234		
R6								
R8								−3.1231

Notes Hypothesis test: $\alpha = 0$

Regions in rows are the dependent variable in the first step of co-integration test; regions in columns are the independent variable in the first step of co-integration test

Critical value at 95 % confidence level is −3.50. Shaded cells indicate that the hypothesis is rejected at the 95 % confidence level, i.e., expenditures in the two regions are co-integrated

change with most of its neighbors. In general, the results indicate that expenditures in the western regions tend to move together, which is consistent with fire activity and expenditures responding to climate patterns that affect a broad geographic area in the western United States (Westerling et al. 2006).

3.3.2 Regional and Temporal Relationships in the SCI Model

3.3.2.1 SCI as a Performance Measure: Comparing Predicted to Actual Expenditure Outcomes

Based on data from fiscal years 1995 to 2004, the SCI model was first used to examine expenditure performance after the FY 2006 fire season. Table 3.4 summarizes expected and actual EPH comparisons for FY 2006 through FY 2012. Since FY 2006, the percentage of fires exceeding expected expenditures has remained relatively stable, between 19 and 25 % exceeding one standard deviation, and between 2 and 6 % exceeding two standard deviations. An exception was in FY 2010, when nearly 40 % of large fires had actual EPH at least one standard deviation above expected, and 19 % exceeding two standard deviations. However, only 78 large fires were evaluated in this year, making it difficult to assess whether the higher percentage of "outliers" represents a significant difference from other years.

Examining EPH performance across regions does not show that any particular region has substantially deviated in recent years from previous expenditure structures when comparing actual to expected EPH. This also suggests that no particular region is driving the national performance rates. Comparing regions across all years, USFS regions had between 13 and 28 % of large fires exceed

Table 3.4 Percent of USFS large fires with actual expenditures per hectare exceeding one or two standard deviations above expected expenditures per hectare, by fiscal year and region

	Number of large fires evaluated	Percent outliers	
		>1 SD	>2 SD
a. Annual fiscal year totals (all regions)			
2006	158	20.9	3.8
2007	164	18.9	3.7
2008	170	24.1	5.9
2009	127	22	1.6
2010	78	39.7	19.2
2011	178	20.8	5.6
2012	192	24.5	3.1
b. Regional totals (all years)			
Northern	127	13.4	2.4
Rocky Mountain	70	28.6	12.9
Southwest	202	26.7	5.4
Great Basin	165	24.2	3.6
California	210	26.2	6.2
Northwest	96	18.8	3.1
Southern	167	22.8	5.4
Northern	30	20	3.3
Total (all years and regions)	1,067	23.2	5.2

expected EPH by more than one standard deviation, and between 2 and 13 % exceed expected EPH by more than two standard deviations. The Rocky Mountain region stands out with the largest share of large fires greater than two standard deviations above expected EPH. However, the small number of large fires evaluated in each region can mean that a few fires in any given year that exceed performance thresholds can result in large changes in percentages.

In summary, there is little evidence to suggest a trend over time in the per-unit expenditures (EPH) of managing USFS fires. Addressing the first research question in this chapter, expenditure performance does not appear to be improving over the time period that the SCI has been used as an evaluation tool. Further, no particular region stands out as a high or low performer on an EPH basis. It is possible that significant variation in fire activity from one year to the next and between regions masks temporal or regional trends. Observations from additional years may help identify these trends in the future.

3.3.2.2 Regression Decomposition of SCI Model Parameters

The basic regression results of EPH on fire size, landscape characteristics, and categorical variables are listed in Table 3.5. As has been found in previous studies, larger fires tend to have lower expenditures per hectare, fires burning in grass tend to

Table 3.5 Expenditure per hectare regression coefficients for pooled and interaction models

	Pooled model (no interactions)	High-cost geog. area interactions	High-cost year interactions
Variable	*Population coefficients*		
LNHECTARE	**−0.329** (0.013)	**−0.346** (0.014)	**−0.318** (0.016)
ERC	**0.011** (9.9e-5)	**0.011** (0.001)	**0.010** (0.001)
BRUSH_FMD	**0.379** (0.047)	**0.348** (0.052)	**0.303** (0.059)
BRUSH4	**0.747** (0.107)	**0.765** (0.119)	**0.881** (0.134)
TIMBER	**0.608** (0.054)	**0.547** (0.060)	**0.587** (0.068)
LNELEV	**0.715** (0.031)	**0.725** (0.034)	**0.750** (0.039)
LNTOT20	**0.047** (0.008)	**0.049** (0.009)	**0.044** (0.010)
WILD	**0.606** (0.135)	**0.732** (0.163)	**0.501** (0.167)
LNWILD_DIST	**−0.514** (0.130)	**−0.612** (0.154)	−0.265 (0.173)
LNTOWN_DIST	0.053 (0.034)	0.070 (0.037)	0.031 (0.042)
DOI	**−1.10** (0.054)	**−0.848** (0.061)	**−1.28** (0.070)
MID_COST_GEOG	**0.465** (0.053)	**0.435** (0.054)	**0.463** (0.067)
HIGH_COST_GEOG	**1.10** (0.056)	**3.28** (0.720)	**1.05** (0.070)
MID_COST_YEAR	**0.189** (0.049)	**0.156** (0.055)	**0.188** (0.050)
HIGH_COST_YEAR	**0.311** (0.043)	**0.325** (0.046)	0.523 (0.537)
Constant	**0.659** (0.261)	**0.419** (0.286)	0.637 (0.328)
Variables with interactions		*Interaction term coefficients*	
LNHECTARE		**0.072** (0.030)	−0.040 (0.026)
ERC		−0.003 (0.003)	0.002 (0.002)
BRUSH_FMD		**0.301** (0.124)	**0.219** (0.098)
BRUSH4		−0.031 (0.260)	−0.325 (0.222)
TIMBER		**0.315** (0.139)	0.097 (0.114)
LNELEV		**−0.207** (0.083)	−0.105 (0.065)
LNTOT20		**−0.042** (0.021)	0.005 (0.017)
WILD		**−0.574** (0.284)	0.103 (0.285)
LNWILD_DIST		0.532 (0.280)	−0.413 (0.264)
LNTOWN_DIST		−0.063 (0.085)	0.067 (0.071)
DOI		**−1.11** (0.129)	**0.439** (0.110)
MID_COST_GEOG		–	0.001 (0.108)
HIGH_COST_GEOG		–	0.080 (0.119)
MID_COST_YEAR		0.069 (0.121)	–
HIGH_COST_YEAR		0.033 (0.115)	–
Log likelihood	−10,026.7	−9,956.6	−10,002.1
R^2	0.388	0.402	0.393
LR stat (d.f. = 13)		140.3	49.3

Obs. = 5,697. Standard errors in parentheses. Coefficient estimates in boldface are statistically significant at the 95 % level

be the least expensive, and those burning in timber tend to be the most expensive. Fires that ignite within a designated wilderness area have higher EPH, but EPH is inversely related to the distance of the ignition point from the wilderness area boundary. Finally, fires managed by the DOI have lower EPH than USFS fires.

The decomposition analysis proceeds by estimating Eq. (3.6) with the high-cost geographic areas category as the C variable, and again with the high-cost years category as the C variable. In both instances there is evidence that the expenditure structure is different for the high-cost categories. Likelihood ratio tests reject the joint hypothesis that all of the interactions are not significantly different from zero (i.e., that the pooled model without interactions is the appropriate model). This suggests that a portion of expenditure differences is related to structural differences in how fires are managed in high-cost geographic areas and in high-cost years.

For high-cost geographic areas, interaction coefficients are significant and positive for fire size (LNHECTARE) and distance to wilderness boundaries. This means that the negative effect of increases in fire size and greater distances to wilderness boundaries are smaller for the high-cost regions, which increases expenditures in those regions. Also, fires that burn in brush/southeast understory pine (known as Fuel Model "D" or FMD) and timber are significantly more expensive for the high-cost regions.

Not all interactions with geographic regions are positive, indicating lower expenditures per hectare associated with some fire characteristics in high-cost regions. For example, the increase in expenditures normally associated with greater housing value within 20 miles of ignition (LNTOT20) is eliminated in high-cost regions. This may suggest that expenditures in high-cost regions are not sensitive to variations in housing values. That is, in high-cost regions (particularly California), fires may be more likely to have high expenditures regardless of property values at risk.

Although the coefficient decomposition for high-cost areas helps to describe the structural differences in how expenditures are determined, it also does not fully explain regional expenditure differences. Indeed, the categorical variable coefficient for high-cost areas is now significantly larger than in the pooled model. This suggests that there are still unobserved characteristics of fires in California and the Northwest that are positively associated with EPH, and that the effect of these factors (and the underlying differences between regions) is larger than previously thought.

A possible explanation for this result is that the combined effect on EPH of the interaction terms for high-cost areas is negative on average. That is, ignoring the different expenditure structures results in underestimates of the differences between the high-cost areas and other areas. Ignoring interactions can bias conclusions about how fire characteristics are related to expenditures per hectare. For example, the relationship between EPH and housing value (LNTOT20) is larger for all regions after controlling for the high-cost area interaction; ignoring the offsetting relationship with housing value in high-cost areas produces a downward bias on the LNTOT20 coefficient in the pooled regression.

The coefficient decompositions for high-cost years yield a more straightforward interpretation of the expanded expenditure structure. The only statistically significant interaction terms are those for the brush/FMD fuel models and the DOI categorical variables, both of which are positive. This suggests that in the high-cost years (2004, 2010–2012) fires with these characteristics are managed differently, and more expensively, than in other years. Further, incorporating the interactions appears to fully explain the expenditure differences between high-cost and other years. The remaining intercept term for high-cost years is now statistically indistinguishable from zero.

Although the decomposed model can identify the source of the expenditure differences between high-cost and other years, it cannot distinguish the reason these differences exist. It is not clear why fires managed by DOI were more expensive in some years (and the past three years in particular), or why fires that ignite in certain fuels would be more expensive to manage in certain years. A topic for future research remains to identify unobserved characteristics that are positively correlated with EPH and are more likely to occur for DOI fires and brush/FMD ignitions in certain years. Similarly, the reason for different expenditure responses to certain variables by region is as yet unexplained, and requires more detailed investigation in the future.

3.3.2.3 Summary

The decomposition of SCI-type regressions for suppression expenditures per hectare suggests two general conclusions: First, expenditure models appear to be more complicated than previously thought, with the structure of how fires are managed from an expenditure perspective varying geographically and over time. Second, significant work remains to identify the underlying causes of these differences. Between regions, incorporating coefficient interactions actually increased the magnitude of unexplained expenditure differences between regions. For annual differences, factors that drive differences in the expenditure structure in particular years are difficult to identify.

3.4 Conclusion

Though this chapter has brought up as many questions as it answered, it has shed additional light on some important questions that need to be answered to understand the rising expenditures on wildland fire. Results indicate that the general overall trend in aggregate suppression expenditures may largely be due to factors outside the control of the land management agencies. Our time series analysis of annual aggregate suppression expenditures is consistent with the observation that expenditures have indeed risen over time. However, year-over-year expenditures follow an unpredictable path that does not adhere to a consistent time trend. Further analysis indicates that expenditures in the western regions tend to move together, which is consistent with fire activity and expenditures responding to climate patterns that affect a broad geographic area in the western United States. Only Region 2 expenditures seem to move independently of the rest of the western Regions.

However, results also show that when looking at the expenditures on individual fires, differences among regions are still largely unexplainable. The analysis of per-fire expenditures indicated that, after taking into account the interaction of high-cost regions with other fire characteristics, differences in expenditures between high-cost regions (Northwest and Pacific Southwest) and the lower-cost regions are actually amplified. This leads to more questions about what unobservable characteristics (that differ between the high-cost and low-cost regions) are not being explained by the model. Conversely, for high-cost years, incorporating the interactions appears to fully explain the expenditure differences between high-cost and other years. However the interaction terms show that in high-cost years, fires managed by the DOI and fires burning in a fuel model of brush/FMD are more expensive. The question then becomes, why?

Determining the answers to these questions is becoming increasingly important. The increasing trend in federal expenditures experienced by land management agencies in recent decades, coupled with declining federal budgets, magnifies the need to determine why wildfire suppression expenditures are increasing. If reducing suppression expenditures or slowing the increase in expenditures is a high priority, knowing the reasons for increasing expenditures can help land management agencies focus expenditure-containment efforts in ways likely to be most effective. If the increases are largely due to factors outside the control of the agencies (like climate and weather), management efforts may be more focused on when and where to fight fires, rather than on hazardous fuel manipulation or specific strategies and tactics. If increasing expenditures are due to changes in vegetation (like increased fuel loads, insect infestations, etc.), then some type of vegetative manipulation may also be helpful. However, if increasing expenditures are more likely due to human factors such as social/political pressures, risk aversion leading to overuse of resources, or increasing populations in the WUI, then the answer may lie more in how to change the behavior and incentives facing fire managers and landowners than in changing the biophysical environment.

When looking at the SCI analyses that have been required of agencies over the past several years (number of fires exceeding expected expenditures by more than 1 or 2 standard deviations), results show little change over time. This could indicate that measures taken to contain expenditures in the past decade have been tempered by other factors affecting expenditures. More research is needed to understand the complex relationships between expenditures, fire characteristics, climate and weather, and human factors to determine the best way to deal with rising suppression expenditures.

Chapter 4
Modeling Fire Expenditures with Spatially Descriptive Data

Abstract A regression model of suppression expenditures based on spatially descriptive fire characteristics in the United States Forest Service, Northern Region (Region 1) is expanded to include all six regions of the western United States. Spatially descriptive landscape and fire characteristics are calculated using available final burned area perimeters for large wildfires (greater than approx. 121 hectares). These characteristics are used as independent variables to describe variations in total suppression expenditures for 419 fires. Hierarchical partitioning is employed to develop a parsimonious regression model, and results are checked for spatial autocorrelation. Results suggest that spatially descriptive data is useful for explaining variations in suppression expenditures, and spatial data with regional controls can account for spatial error patterns observed in the dependent variable. Spatially descriptive models have the potential to be used for a variety of applications where expenditure estimates are needed and planning and management activities rely on spatially explicit information that can be used in expenditure models. However, finer-scale geospatial data is necessary to integrate spatially descriptive expenditure models with spatially explicit fire management planning tools (such as outputs from fire simulations).

Keywords Regression expenditure models · Spatial autocorrelation · Hierarchical partitioning · Spatial characteristics · Fire perimeters

4.1 Introduction

In their Region 1 model, Liang et al. (2008) identified two primary factors that determine expenditures on wildfire suppression efforts: fire size and the share of a fire's burned area that was privately owned. Fires with larger size or with more private land area in their perimeter were found to be more expensive to control in the Northern Region. In contrast to previous studies that examined the role of fire characteristics based on the ignition point, Liang et al. (2008) calculated fire characteristics for the entire burned area of each fire observation. A parsimonious model using relatively few explanatory variables was developed to explain a large share of sample variation in suppression expenditures. The results suggest that the spatial composition of landscape characteristics and how managers respond to different spatial arrangements of landscape characteristics are important determinants of expenditures.

In this chapter we explore relationships between spatially explicit landscape characteristics and suppression expenditures in a larger spatial scale. Using the approaches of Liang et al. (2008) as a guide, which examined expenditures for 100 fires in the Northern Region (which includes Montana and parts of northern Idaho, North Dakota, and South Dakota), we specify a model of suppression expenditures for the entire western United States based on spatial characteristics within each fire's burned area. Data are drawn from fires with "ignition point" data available, similar to data used in the SCI model (Gebert et al. 2007). Landscape characteristics, such as fuel types, land ownership, and protected status, are calculated based on the area within the final fire perimeter.

The expenditure model for the western United States can indicate whether the findings based on a single region—the Northern Region—are applicable to other regions or the nation as a whole. This analysis is aimed at determining whether spatially descriptive characteristics of fires have a consistent impact on expenditures across the western United States or whether region-specific effects dominate. Further, the results can be compared to comparable ignition point models to assess whether more detailed spatial and geographic data will improve the performance of expenditure models. These results can have implications for decision support tools, such as the Wildland Fire Decision Support System (WFDSS, Calkin et al. 2011b) that include predicted expenditure modules, post-season evaluation of expenditure performance, and the evaluation of how landscape changes (such as fuel treatments) may affect suppression expenditures of future fires.

4.2 Insights from the Region 1 Model

In the Region 1 model, Liang et al. (2008) studied suppression expenditures on large wildland fire spent by the USFS. Among 16 potential characteristics representing fire size and shape, private properties, public land attributes, forest and fuel conditions, and geographic settings, only fire size and private land had a significant influence on suppression expenditures. A parsimonious model to predict suppression expenditures was developed with hierarchical portioning (Chevan and Sutherland 1991), which shows that fire size and percentage of private ownership within the fire perimeter explained 58 % of variation in expenditures.

With the average fire size (925 ha), suppression expenditures dramatically increased as the proportion of private land within the burned area increased from 0 to 20 percent. Suppression expenditures peaked at around $3 million with 20 % of private land. As this percentage continued to increase, suppression expenditures started to slowly decline and stabilized in the neighborhood of $1 million. With the average percentage of private land within burned area (10 %), suppression expenditures increased monotonically from around $280,000 to $28 million, as fire expanded in size from 148 to 22,000 ha. The results suggested that efforts to contain federal suppression expenditures need to focus on the highly complex, politically sensitive topic of wildfires that transition between federal and private lands. Detailed descriptions of data, methods, and findings can be found in Liang et al. (2008).

4.3   A Spatial Expenditure Model for the Western United States

Expanding the spatial expenditure model to include all of the western United States can provide additional insights about suppression expenditures. Conclusions from a single region may not apply to a wider geographic scope due to spatial heterogeneity and regional differences in strategic wildfire management. The expanded model is developed in a similar manner as the Region 1 model discussed above to facilitate comparisons of conclusions and insights with the original model. For broader applicability, some variables differ from the Region 1 application. These include several of the forest and fuel conditions variables from the Region 1 model (e.g., surface-to-area volume, packing ratio, moisture content, rate of spread, flame length, fine fuel load) that have been replaced by spatial fuel model and energy release component (ERC) variables.

4.3.1 Data

Observations are drawn from USFS large fires (with final burned area greater than 121 hectares) from fiscal years 2006—2011 that have final burned area perimeters available. Individual fire records are drawn from NIFMID. Fire perimeters gathered from the National Interagency Fire Center (NIFC) incident FTP service were overlaid with several geospatial data layers to calculate geospatial characteristics for each fire (see Table 4.1 for data sources).

Records of fires that were part of a fire complex, that is, two or more individual fires managed as a single incident, are dropped from this analysis because it is not possible to accurately apportion total expenditures on the complex to the individual component fires. The final calibration dataset is comprised of spatially explicit attributes of 419 fires in the western region (Table 4.1).

The response variable in this analysis is the natural log of total USFS suppression expenditures (*lnfs_exp*), adjusted for inflation to 2012 dollars. Expenditure data is drawn from USFS financial databases and NIFMID, which tracks financial data for fires where Federal agencies have the primary management

Table 4.1 Description of variables studied in this chapter (obs = 419). Variable names in bold are selected from preliminary regressions to be included as candidate variables for the hierarchical partitioning analysis

Variable	Description	Source	Mean	SD
lnfs_exp	Natural log of USFS expenditures, in 2012 $ (Dep. Var.)	FFIS	13.8	1.99
lngis_ha	Natural log of final burned area in hectares, calculated from final GIS perimeter	NIFC FTP	8.09	1.48
one_day	Binary indicator of one-day duration (based on fire occurrence discovery and strategy-met dates)	NIFMID	0.062	0.242
erc_max	Maximum relative energy release component value between discovery date and strategy-met date within the final burned area	Calculation based on Abatzoglou (2013) data	92.7	13.2
erc_std	Standard deviation of relative energy release component values between discovery date and strategy-met date within final burned area	Calculation based on Abatzoglou (2013) data	14.1	11.4
ln_avelev	Natural log of the mean elevation (in ft.) within the final burned area	LANDFIRE	7.37	0.442
Special designated areas		WFDSS		
wild_burn	Binary indicator of whether the fire burned any area within a designated wilderness area		0.317	0.466
wild_sh	Share of final burned area that was within a designated wilderness area		0.202	0.359
ira_burn	Binary indicator of whether the fire burned any area within an inventoried roadless area		0.487	0.500
ira_sh	Share of final burned area that was within an inventoried roadless area		0.223	0.331
slp_1	Share of final burned area with slope class 1 (<20 %)	LANDFIRE	0.374	0.304
usfs_sh	Share of final burned area in USFS ownership	WFDSS	0.818	0.293
timber_spa	Share of final burned area with timber fuel models (FBFM8, FBFM9, FBFM10)	LANDFIRE	0.378	0.300
Housing value		U.S. Census		
lnhouse8_in	Natural log of value of housing (000's of $) between fire perimeter and 8-km (5-mile) radius of final perimeter		4.66	5.88
lnhouse16_8	Natural log of value of housing (000's of $) between 8- and 16-km radius of perimeter		7.14	6.05
lnhouse32_16	Natural log of value of housing (000's of $) between 16- and 32-km radius of fire perimeter		11.4	4.32

(continued)

Table 4.1   (continued)

Variable	Description	Source	Mean	SD
Aspect classes		LANDFIRE		
asp_123	Share of final burned area in N, NE, and E aspects		0.362	0.181
asp_456	Share of final burned area in SE, S, or SW aspect		0.397	0.184
USFS regions		NIFMID		
reg_1	Binary indicator for Northern region (reference category)		0.105	0.307
reg_2	Binary indicator for rocky mountain region		0.069	0.254
reg_3	Binary indicator for Southwest region		0.279	0.449
reg_4	Binary indicator for great plains region		0.193	0.395
reg_5	Binary indicator for California region		0.239	0.427
reg_6	Binary indicator for Northwest region		0.115	0.319

Data sources FFIS Foundation Financial Information System, which is being replaced by the Financial Management Modernization Initiative (*FMMI*), available at http://info.fmmi.usda.gov/, accessed 9/3/2013; *NIFMID* National Interagency Fire Management Integrated Database, maintained at the USDA National Information Technology Center in Kansas City, MO; NIFC FTP—available at ftp://ftp.nifc.gov/Incident_Specific_Data/, accessed 7/24/2013; *WFDSS* Wildland Fire Decision Support System databases available at http://wfdss.usgs.gov/wfdss/WFDSS_Data_Downloads.shtml, accessed 7/24/2013; LANDFIRE—version 1.2.0 available at http://www.landfire.gov/lf_120.php, accessed 7/24/2013

responsibility. All of the fires included in the estimation data set were primarily managed by the USFS, and the USFS expenditures account for about 88 percent of total expenditures on these incidents.

Potential explanatory variables were developed to provide a more detailed spatial description of variables previously either reported or calculated from the ignition point in the original SCI model. The variables describing spatial characteristics fall into three categories: summary variables, composition variables, and distribution variables. Summary variables represent a value (usually the mean) that summarizes a given characteristic over the spatial extent of a fire's burned area (e.g., average elevation). Composition variables describe the portion of the burned area that fall into different categories of a characteristic (e.g., slope classes, or land ownership categories). Distribution variables describe how a characteristic is distributed across the burned area, or how a characteristic varies in space (e.g., aggregate housing value at different distances from the burned area).

The Energy Release Component (ERC) is an approximation of dryness of the U.S. National Fire Danger Rating System (Deeming et al. 1977) calculated from a suite of meteorological and site specific variables. ERC is used by U.S. federal land agencies both operationally (e.g. Predictive Services), in simulation models that predict fire size and probability (Finney et al. 2011a, b), as well as fire expenditure estimation (Gebert et al. 2007). Within the model, ERC was calculated using two variables: The maximum daily ERC value over the course of the fire (*erc_max*), and the standard deviation of ERC during the fire (*erc_std*). ERC can vary over time during a fire depending on changing weather conditions. Many fires in the estimation sample had periods of moderate ERC values, reflecting weather conditions that may be favorable for suppression activities. Quiescent periods of weather have been shown to be correlated with increased likelihood of fire containment (Finney et al. 2009), which can affect expenditures by shortening fire duration (Gebert and Black 2012).

4.3.2 Methods

The expanded spatial model largely follows the modeling approach set out in Liang et al. (2008) for Region 1. We begin with preliminary ordinary least squares (OLS) regressions of total USFS expenditures on spatially explicit landscape characteristics and other controls (like regional identifiers). Candidate explanatory variables were identified from the preliminary regressions based on statistical significance and insights from previous research. We then conduct a hierarchical partitioning (HP) analysis to identify a subset of variables that account for the greatest share of variance in suppression expenditures. The resulting parsimonious regression models are evaluated for spatial autocorrelation.

Hierarchical partitioning (Chevan and Sutherland 1991) provided key insights into expenditure structures for Region 1, including identifying the independent and combined contribution of fire size, private land and developed areas, and other landscape characteristics to the goodness-of-fit of suppression expenditure models. Prior to conducting hierarchical partitioning, preliminary regression models were used to identify a list of 12 candidate variables that were significantly related to expenditures at the 95 % level or higher (Table 4.1). The hierarchical partitioning regressions and analysis were executed using the hier.part package (Walsh et al. 2003) in the R system (R Core Team 2013). Based on the hierarchical portioning, the explanatory variables with an independent contribution to the overall goodness-of-fit greater than 1/12 (8.33 %) were selected to compose the final parsimonious model (Table 4.2).

4.3.3 Results

Hierarchical partitioning candidate variables are listed in Table 4.2 along with their independent contribution to goodness-of-fit (R^2). As in the Region 1 model, fire size (*lngis_ha*) accounts for the greatest share of goodness-of-fit at about 22 %. Average ERC during the fire contributes about 11.8 %, indicating that variation in weather and fuel conditions helps explain expenditure variations between fires. The Region 5 (California) dummy variable and the housing value variables also contribute around 10 % to goodness-of-fit, whereas no other variable contributes

Table 4.2 Percentage independent contribution to the overall goodness-of-fit (R^2) calculated by hierarchical partitioning. (Dep. Var. = lnfs_exp; obs. = 419). Variables in bold which have independent contribution greater than 1/12 (8.33 %) were selected to compose the parsimonious model

Variable	Independent contribution (%)
lngis_ha	**22.31**
one_day	5.61
erc_max	**11.79**
erc_std	4.973
ln_avelev	5.09
wild_sh	4.29
usfs_sh	4.90
timber_sh	6.00
lnhouse8_in	**9.50**
lnhouse16_8	**8.49**
reg_5	**10.74**
reg_6	6.31

more than 8 %. In total, fire size, average ERC, the Region 5 dummy, and the housing value variables account for about 63 % of overall goodness-of-fit.

It is noteworthy that private land variables were not candidate explanatory variables, whereas in the original Region1 model (Liang et al. 2008) private land was associated with higher expenditures and contributed about 11 % to goodness-of-fit. Initial regressions indicated that private land was not significantly related to expenditures. The only jurisdiction category that was significantly related to expenditures was the share of burned area under USFS ownership (which tends to be positively related to expenditures).

Regression estimates are presented in Table 4.3 for the full model and the parsimonious model where variables that contribute less than 8.3 % to goodness-of-fit have been dropped. Overall the models perform similarly to the Region 1 models, with the parsimonious model explaining about 52 % of variation in expenditures, and the full model explaining about 66 %. All of the coefficients in both models are highly significant and of the expected sign.

Fire size is positively associated with expenditures, similar to previous findings. The magnitude of the coefficient is on the low end of the range reported in the Region 1 model Liang et al. (2008). In the expanded parsimonious model, the coefficient implies that for the average fire a 10 % increase in fire size results in a 6.5 % increase in expenditures.

Maximum ERC is positively associated with expenditures in both the full and parsimonious models, indicating that more extreme weather and fuel conditions increase expenditures. ERC was not a variable included in the Region 1 model, although the Region 1 model controlled for several forest and fuel condition variables. These variables individually (or as a group) did not contribute greatly to the explanatory power of the Region 1 model. In the expanded model, it may be that the calculation of ERC summarizes forest, fuel, and weather conditions in a way that is more closely related to fire behavior (e.g., how intense a fire is likely to burn) and fire management (e.g., how difficult it is to suppress a fire).

Table 4.3 USFS fire expenditure models for the Western regions only (Dependent Variable = lnfs_exp; number of observations = 419)

Variable	Full model		Parsimonious model	
	Coeff.	Robust S. E.	Coeff.	Robust S. E.
lngis_ha	0.705	0.041	0.646	0.046
one_day	−0.684	0.301		
erc_max	0.037	0.006	0.041	0.006
erc_std	−0.014	0.006		
ln_avelev	0.396	0.189		
wild_sh	−0.992	0.183		
usfs_sh	1.34	0.262		
timber_sh	1.01	0.234		
lnhouse8_in	0.032	0.012	0.032	0.014
lnhouse16_8	0.038	0.012	0.035	0.013
reg_5	1.92	0.178	1.28	0.152
reg_6	1.55	0.207		
constant	−0.307	1.45	4.12	0.575
R^2	0.661		0.518	
RMSE	1.1734		1.386	

Although the standard deviation in ERC was dropped from the parsimonious model, in the full model greater variation in ERC during the fire was significantly associated with lower expenditures. This finding is consistent with periods of quiescent and moderate conditions providing opportunities to contain a fire earlier and reduce expenditures (Finney et al. 2009).

Housing values are a significant predictor of expenditures in the expanded models. Greater total home values in proximity to the fire are associated with greater expenditures. However, there does not appear to be a distance gradient for the effect of housing values on expenditures. The coefficients for the two housing value variables (*lnhouse8_in*, *lnhouse16_8*) are roughly equal in both models, suggesting that housing value at risk within 8 km of the final burned area has the same effect as that between 8 and 16 km from the perimeter.

The housing value results are in contrast to the Region 1 results, in which housing value and wildland-urban interface land were not significantly associated with expenditures after controlling for fire size. This result likely indicates differences in the expenditure structure for the other regions compared to Region 1. The average housing value within 8 km of the fire perimeter is roughly equal in the Region 1 sample and the expanded sample (about $100,000); the different results may be due to regional differences in the distribution of housing values within the 8 km buffer, or regional differences in how managers respond to threatened structures or unoccupied private timber and range lands.

An important result from Liang et al. (2008) was that after controlling for spatial characteristics in the model, residuals did not exhibit spatial autocorrelation. This suggests that the model independent variables are capable of adequately controlling for spatial relationships among individual fire observations, at least for fires in the Northern Region. In the western U.S. model, greater geographic scope and spatial heterogeneity among observations could introduce spatial autocorrelation.

Spatial autocorrelation among sample observations was estimated using a nonparametric spatial correlogram function (Bjørnstad and Falck 2001). This method incorporates a smoothing spline to measure the correlation between values of the dependent variable or model residuals from sample pairs of observations over a continuous distance function, without assuming any functional form a priori. A bootstrap approach with 500 replications was used to determine the 95 % confidence intervals (i.e., the 2.5 and 97.5 % quantiles) of the mean spatial correlogram distribution (Efron and Tibshirani 1993).

Prior to estimating the fire expenditure models, the dependent variable *lnfs_exp* exhibits spatial autocorrelation, as the mean spatial correlogram was significantly above zero when the distance between pairs of sample was less than 400 km (Fig. 4.1a). In contrast, the fully specified and parsimonious (HP) models were able to control for the spatial autocorrelation. This finding is likely due to the inclusion of regional dummy variables in the fully specified and parsimonious models, which explain a significant amount of spatial variation in expenditures.[1] The residuals from the full model display a spatial correlogram that is not significantly different from zero over its range (Fig. 4.1b), and those from the parsimonious model display a correlogram that is significant only over a small range (<50 m). This finding suggests that residuals from the full and parsimonious models are in general spatially independent. That is, the models were able to control for the spatial autocorrelation in expenditures, making predictions and inferences more reliable.

4.4 Model Interpretation and Applications

The expanded spatial model of suppression expenditures is useful for providing insights on how spatial variation and arrangement of landscape characteristics are related to expenditures. Previous models that use landscape characteristic data based on the ignition point (e.g. Gebert et al. 2007) are well-suited to explaining expenditure differences when the primary source of variation between fires is captured at the ignition point. When these models were developed, spatial descriptions of fire characteristics using final fire perimeters were not widely available, and ignition-point data represented the best available data to describe wildfires. The general hypothesis that has motivated the use of spatially explicit data is that variations in landscape characteristics within a fire, and differences

---

[1] See Chap. 3 in this volume for a more detailed discussion of regional variations in expenditures.

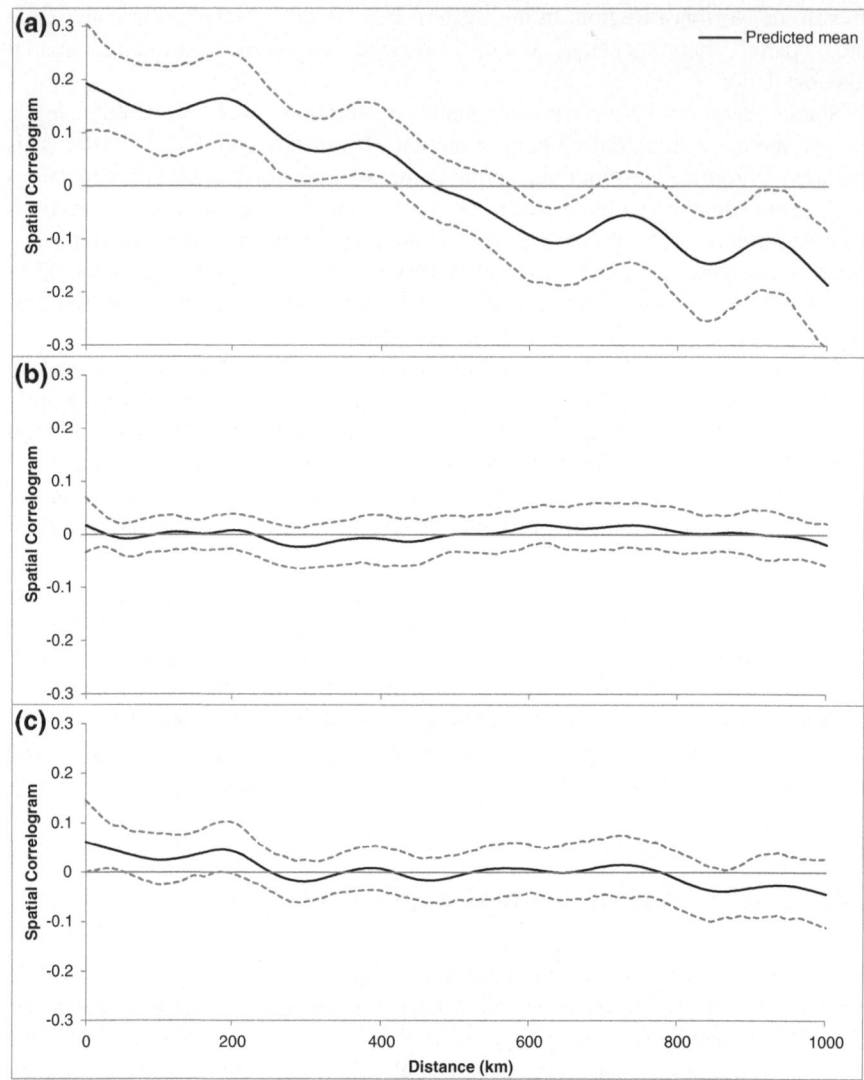

Fig. 4.1 Spatial autocorrelation of **a** response variable values, **b** residuals from the full model, and **c** residuals from the parsimonious model selected by hierarchical partitioning. Solid lines represent mean spatial correlogram, and broken lines represent 95 % confidence

between fires in how these characteristics are arranged within a fire perimeter, can affect how managers respond to fires and thus expenditures.

The fully specified model for the western United States indicates that spatial patterns of landscape characteristics are important determinants of expenditures. For example, the composition of land ownership, fuels, and protection designations (e.g., wilderness areas) within the burned area are significantly associated with expenditures in a way that is not accounted for by ignition-point data. The spatial model allows for conclusions to be made about how marginal changes in composition affect expenditures (e.g., the effect on expenditures of burning additional area in timber fuels), rather than distinctions among categories of fires (e.g., whether or not the fire started in timber fuel types).

The spatial model also lends itself to more nuanced analyses of how policies, forest programs, and suppression strategies may affect expenditures. The ignition point SCI has been used to estimate the potential change in suppression expenditures resulting from a fuel treatment program (Thompson et al. 2013d) and a strategy that promotes allowing wildfires to burn for achieving resource benefits (Houtman et al. 2013). In these applications, the primary factor that affected a change in expenditures was a change in final fire size. The fully specified spatial model implies that for fires of equal size and the same ignition-point fuel, a fire that burns a lower share of timber is likely to be less costly. The use of the spatial model in determining the potential suppression savings from fuel treatment programs or beneficial fire policies is currently limited due to the absence of mapped fuel treatment project boundaries and the coarseness of the current fuel models (i.e. timber, brush, grass). Research is currently underway to explore the influence of fuel treatments and past fires on suppression expenditures using spatially explicit data. This requires spatial fuel treatment and fire history data as well as more refined fuel models to relate changes in landscape characteristics that affect fire behavior to management decisions and, ultimately, fire expenditures.

A spatial model of expenditures may also be useful for improving expenditure estimates in decision support systems. The Wildland Fire Decision Support System (WFDSS) is increasingly incorporating fire simulation models that provide fine-scale geographic information about potential fire activity under varying conditions and strategy choices (Calkin et al. 2011b, d). These tools may improve the efficiency of fire management efforts (Hesseln et al. 2010). The spatial model could provide more details in the expenditure consequences of potential fire outcomes and provide managers with an early warning of when fire growth in certain areas may increase or constrain expenditures.

4.5 Conclusion

The expenditure models originally presented in Liang et al. (2008) represented the first analysis (to our knowledge) of how spatially explicit descriptions of landscape characteristics are related to wildfire suppression expenditures. Although limited in scope to a small set of fires in the Northern Region, the results indicated the potential of spatially descriptive expenditure models to improve wildfire cost estimation. The models presented in this chapter suggest that spatially descriptive expenditure models are also useful for the entire western United States.

There is an increased interest in spatially explicit descriptions of wildfire outcomes, particularly suppression expenditures. Examples of potential uses include forecasting suppression expenditures based on spatial variation in landscape characteristics and the likelihood that fire occurs in different locations (e.g. Preisler et al. 2011), examining the expenditure consequences on fuel treatment programs (Thompson et al. 2013d), developing risk-based suppression budgets (Thompson et al. 2013a), improving our understanding of the implications of climate change on land management budgets, and providing more detailed information to aid in decision support systems [e.g., WFDSS, (Calkin et al. 2011c)]. In all of these cases, the underlying mechanisms and models that drive differences in fire outcomes (and thus expenditures) are inherently spatial in nature. Developing spatially descriptive expenditure models can leverage the trove of spatial data associated with these applications, and provide more informative analyses for managers and decision makers.

Although the spatial expenditure model in this chapter represents an expansion on previous spatial efforts, much research remains to bring spatially descriptive models into more widespread use. First, we have not attempted to compare the performance and predictive ability of the spatial model to the existing ignition-point models. These models (e.g. Gebert et al. 2007) perform reasonably well and can be used in a wide variety of applications; it is not yet known whether spatial models can offer a significant improvement. Second, the spatially descriptive models have thus far relied on relatively coarse-scale geographic data. Future applications would benefit from fine-scale data that can distinguish the impacts on expenditures from landscape and vegetative patterns and the effects of fuel treatments and past fire perimeters that are important for suppression efforts. For example, the data used in this chapter cannot distinguish public–private land intermix and in-holdings, or heterogeneity of vegetation types. Finally, future research may benefit from exploring factors that affect how fire managers commit and use resources during an incident. Suppression expenditures are ultimately a consequence of decisions to order and use personnel and equipment over the course of a fire. Understanding the factors that influence these decisions, including spatial characteristics, may provide a clearer link between expenditures and landscape characteristics.

Chapter 5
Linking Suppression Expenditure Modeling with Large Wildfire Simulation Modeling

Abstract Land management agencies face uncertain tradeoffs regarding investments in preparedness and pre-fire management versus future suppression expenditures and impacts to valued resources and assets. This chapter illustrates one potential method for linking suppression expenditure models with fire simulation models in order to estimate suppression expenditures and evaluate alternative risk mitigation strategies. A case study application in the Deschutes National Forest illustrates how fire simulation outputs can be linked with geospatial information from other databases to calibrate a model of wildfire suppression expenditures. The resulting output can describe the expenditure consequences of different spatial and temporal arrangements of fuel treatments. In the case study example, fuel treatments that reduce median fire sizes on the landscape by 5.25 % yield suppression expenditure savings of 5.03 % over a ten-year time span. Within the treated areas only, effects of fuel treatments on fire size and expenditures are larger as more fires interact with the treated portions of the landscape. The approach illustrated in the case study allows analysts to address a variety of salient wildfire management and policy questions, including comparative assessments of alternative wildfire management strategies and comparisons of expected suppression expenditures across landscapes and geographic areas.

Keywords Fuel treatments · Wildland fire simulation · FSIM · Stratified cost index · Treatment expenditures · Tradeoffs · Deschutes national forest

5.1 Introduction

Land management agencies face uncertain tradeoffs regarding investments in preparedness and pre-fire management versus future suppression expenditures and impacts to valued resources and assets. The expected expenditures on and impacts of fire are not equal across landscapes or across regions, suggesting opportunities exist for efficiency gains by prioritizing risk mitigation investments where net

M. S. Hand et al., *Economics of Wildfire Management*, SpringerBriefs in Fire,
DOI: 10.1007/978-1-4939-0578-2_5, © The Author(s) 2014

wildfire management expenditures and detrimental impacts can be minimized (Butry et al. 2010). Prospective evaluation of likely mitigation effectiveness facilitates analysis of tradeoffs across land management objectives and wildfire management expenditures.

A number of factors are thought to influence suppression expenditures, including incident management strategy, location of the ignition and its proximity to human communities and private property, climatic and weather conditions, fuel types, and fire-related characteristics such as size, behavior, and duration (Calkin et al. 2005; Gebert and Black 2012; Gebert et al. 2007; Gude et al. 2013; Liang et al. 2008; Yoder and Gebert 2012). The estimation of expected suppression expenditures can serve as a performance benchmark against which to evaluate actual expenditures (Gebert et al. 2007; Thompson et al. 2013a) and can help to forecast budgetary demands (Abt et al. 2009; Preisler et al. 2011; Prestemon et al. 2008). Further, a suppression expenditure model can facilitate the assessment of how wildfire management actions may affect, or can be designed to affect, future suppression expenditures (Fitch et al. 2013; Thompson et al. 2013d; Prestemon et al. 2012).

Figure 5.1 presents a conceptual model for analyzing interactions between wildfire management actions, wildfire activity, and suppression expenditures. Fundamentally the process entails modeling first how management actions will affect the extent and intensity of wildfires, and second how these changes will affect suppression expenditures. A key logical element is the linkage of factors influencing suppression expenditures with factors that can be impacted through wildfire management actions. These actions could include pre-fire investments in ignition prevention programs, response capacity, and hazardous fuels reduction, as well as changes in strategic incident response. Insofar as management actions and influencing factors are well represented in wildfire models or can be sufficiently captured with expert judgment, potential impacts to suppression expenditures can be estimated. This basic analytical framework can be brought to bear across geographic (landscape to national) and temporal (single incident to multiple fire seasons) scales.

The likely direction and magnitude of potential suppression expenditure impacts will depend on the wildfire management context and mitigation objectives and may entail the consideration of spatiotemporal tradeoffs. Ignition prevention programs and enhancing response capacity both aim to reduce suppression expenditures by excluding fire from the landscape, although fuel accumulation could lead to increased future wildfire activity. Evaluating the impacts of alternative fuel treatment and incident response strategies is more complicated, owing to the wide range of possible objectives. Strategies oriented towards resource protection objectives and limiting the spread of fire (e.g., Ager et al. 2010) could reduce fire sizes leading to reduced suppression expenditures. The degree to which fuel treatments can measurably influence wildfire extent will depend on the type, spatial pattern, and areal extent of treatments (Collins et al. 2010). To the contrary, strategies oriented towards restoration objectives (e.g., Ager et al. 2013) could seek to promote ecologically beneficial fire on the landscape, focusing more on

Fig. 5.1 Influence diagram of wildfire management actions and their relation to primary factors driving wildfire extent and intensity as well as suppression expenditures. Boxes in *light gray* are management actions, and boxes in *dark grey* are analytical outputs of interest. Figure modified from Calkin et al. (2011a)

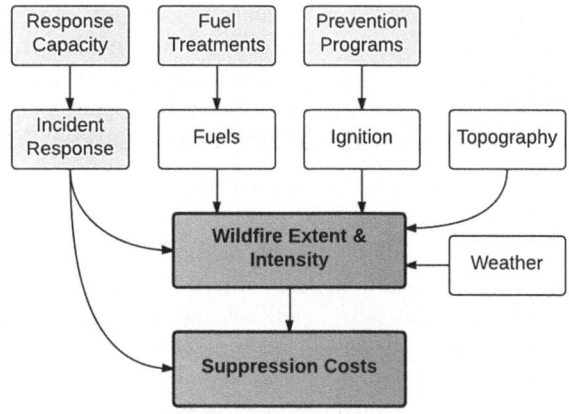

reducing burn severity and subsequent resource damage. Reduced severity could lead to moderated incident response strategies that use significantly fewer fire-fighting resources, although longer incident durations could still lead to high expenditures (Gebert and Black 2012).

This chapter illustrates one potential method for linking suppression expenditure models with fire simulation models in order to estimate suppression expenditures and evaluate alternative risk mitigation strategies. First, the modeling framework is described, as are the specific wildfire and suppression expenditure models that we use. The modeling framework we present is most suitable for evaluating management actions and strategies that aim to reduce the occurrence and areal extent of large wildfires, as opposed to those that aim to restore wildfire to fire-adapted ecosystems. The fire modeling component is similar to the work of Cochrane et al. (2012) who used simulation modeling to explore how previously implemented fuel treatments may have influenced final fire sizes. Second, we offer results from a case study on the Deschutes National Forest in Oregon, United States (Thompson et al. 2013d), examining scenarios that vary landscape fuel conditions and the frequency of large wildfire occurrence. Lastly we discuss challenges and opportunities for future applications linking wildfire simulation and suppression expenditure models.

5.2   Wildfire and Suppression Expenditure Modeling

The foundation of the wildfire and suppression expenditure modeling approach is the simulation of escaped large wildfire occurrence, growth, and containment across the planning scale of interest. Wildfire simulation outputs can then be assimilated with geospatial information from other databases and fed into the wildfire suppression expenditure model. It is important to note that the choices and assumptions behind modeling the effects of management actions on wildfire activity may exert a significant influence on suppression expenditure results. For instance, changes to fire behavior fuel models and canopy characteristics are particularly important when evaluating fuel treatment strategies.

A stylized process for estimating suppression expenditures under various wildfire management scenarios follows five steps:

1. Obtain or generate spatial data for modeling wildfire on the reference landscape.
2. Simulate wildfire under current landscape conditions and management policies (existing conditions simulation).
3. Update landscape conditions and/or management policies appropriately to reflect mitigation strategies; re-simulate wildfire (post-treatment simulation).
4. Feed fire modeling outputs and other data into suppression expenditure model.
5. Compare expected suppression expenditures under existing and post-treatment conditions.

Figure 5.2 depicts the major modeling inputs and outputs, and their interactions in this modeling framework. The specific models we use were developed and are currently used by the USFS and DOI agencies: FSIM, a stochastic large wildfire modeling system (Finney et al. 2011b); and SCI, a suppression expenditure regression model (Gebert et al. 2007). Both FSIM and SCI are intended for modeling "large" wildfires, typically defined as greater than 121 ha in size. FSIM combines sub-models for fire weather, occurrence, spread, and containment in order to estimate wildfire extent and intensity across thousands of simulated fire seasons. Each simulated season can result in zero, one, or many large wildfires, depending on historical fire occurrence patterns and the extent of the simulated landscape. Key spatial inputs to FSIM are landscape variables relating to fuels and terrain conditions (surface fuel model, canopy cover, canopy height, canopy base height, canopy bulk density, elevation, slope, and aspect). In addition to outputting information on individual fire sizes and perimeters, FSIM also aggregates simulation results to quantify annualized burn probabilities, which are crucial for complementary risk-based assessments (Thompson et al. 2013c; Scott et al. 2012).

The coupling of FSIM with SCI enables an ensemble approach to expenditure estimation that accounts for uncertainty surrounding the conditions driving wildfire occurrence and spread. Aggregating expenditures on a seasonal basis allows for exploration of the range of variation of possible suppression expenditures for particularly active fire seasons. These two models are especially well suited to evaluate suppression expenditure impacts of hazards fuels treatments. FSIM controls make it possible to hold the number and location of ignitions, fire weather conditions, and suppression effort constant across different simulations. This allows for the effects of fuel treatments to be isolated (apart from a degree of stochasticity induced via random spotting); post-treatment changes in expenditures are driven primarily by fuel-treatment related changes in fire size.

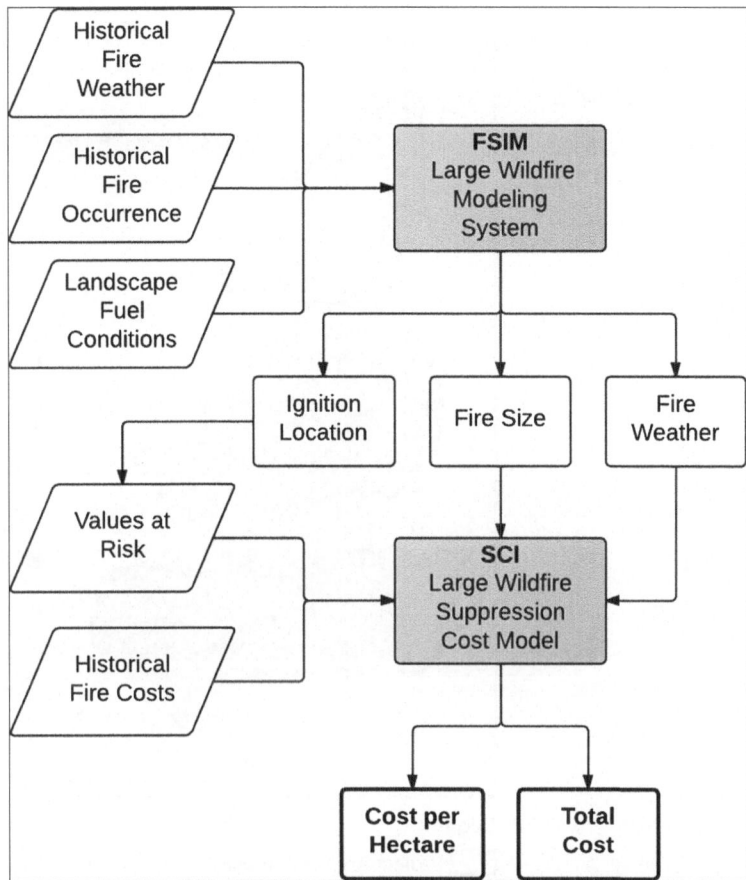

Fig. 5.2 Flowchart of primary wildfire and suppression expenditure modeling components

5.2.1 Case Study: Deschutes National Forest, Oregon, U.S.A

As a case study we turn to an analysis performed on the Deschutes National Forest in Oregon, U.S.A, the details of which are presented in Thompson et al. (2013d). We first illustrate the use of FSIM and SCI to estimate likely suppression expenditures given current landscape conditions, using results from 10,000 simulated fire seasons. We then illustrate how we modeled potential suppression expenditure impacts associated with different wildfire management scenarios.

Figure 5.3 presents a map of the fire modeling landscape, comprising about 209,000 hectares. The outlined area represents the boundaries of the Deschutes Collaborative Forest Project (DCFP), identified as a priority landscape for wildfire risk mitigation. The fire modeling landscape includes a buffer around the DCFP study area in order to account for remote ignitions that could grow large and burn

Fig. 5.3 Deschutes
collaborative forest project
(DCFP) study location, with
areas of implemented and/or
proposed fuel treatments
highlighted. Figure from
Thompson et al. (2013d)

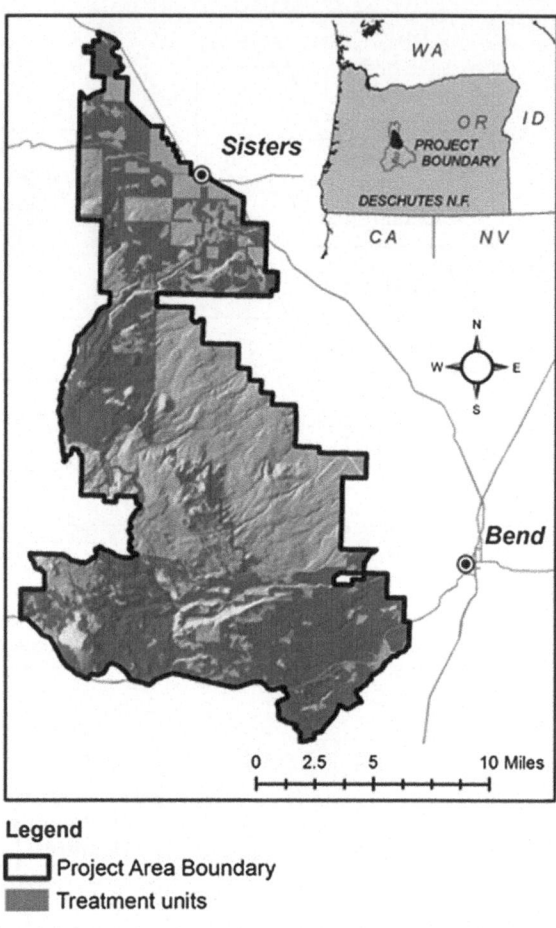

Legend
☐ Project Area Boundary
▪ Treatment units

into the project area and into treated locations. Areas highlighted in grey in the
map represent proposed or implemented fuel treatment locations, many of which
are proximal to the communities of Sisters and Bend.

Nearly half of the DCFP landscape is projected to receive fuel treatment during
the planning period from 2010 to 2019. Forest staff provided spatial data on
vegetation and fuel layers for existing conditions (EC), and additionally provided
information on treatment type, treatment polygons, and expected post-treatment
(PT) fuel conditions. For modeling purposes we used a single fire modeling
landscape to capture the cumulative effect of all planned treatments, based on the
assumption that the effective lifespan of treatments extends across the analysis
period. Running FSIM and SCI on the EC landscape quantifies expected suppression expenditures absent pre-fire management intervention. Comparing
expected suppression expenditures on the EC and PT landscapes depicts expected
expenditure impacts associated with the hazardous fuels reduction treatments.

A secondary analysis considers how suppression expenditures might change with varying frequencies of large wildfire occurrence. This scenario could reflect varying levels of investment in ignition prevention programs or initial attack capacity. Further, the scenario could serve as a forecast of potential future expenditures under a changing climate with increased wildfire activity (although, at present we do not model vegetation or fuel changes associated with climatic variables). The scenarios we consider model 1, 3, or 5 escaped large wildfires per year. Using Monte Carlo sampling techniques we can estimate the variability in total suppression expenditures surrounding any given number of escapes. As described earlier this aggregated analysis can help identify just how extreme and high-cost fire seasons could become.

5.2.2   Case Study Results

Simulation results for wildfire extent and suppression expenditures tend to align well with historical observations and expectations. Figure 5.4 graphs suppression expenditures against fire size for all simulated fires on the EC landscape, showing a positive correlation between size and expenditures. The substantial variance in expenditures for a given fire size is indicative of the influence of other locational variables such as distance to a town, proximal housing value, and fuel type. Due to a limited historical record within the study area itself we compared simulation results to large wildfire data for the entire Deschutes National Forest, over the years 2000–2011. A consequence is the inclusion of historical wildfires that ignited in remote areas where suppression expenditures are generally lower. Across the fire modeling landscape the simulated mean and median fire sizes were 3,855 and 1,192 hectares, respectively, compared to historical values of 3,905 and 870 hectares. Simulated mean and median suppression expenditures were $8,990,166 and $5,071,995, respectively, compared to historical values of $6,169,476 and $2,876,921. Again, the higher simulated results are due at least in part to the proximity of the DCFP to the communities of Sisters and Bend, and the positive influence of housing value on suppression expenditures.

The first management scenario explores the impact of landscape-scale fuel treatments by comparing simulation results on the EC and PT landscapes. Across the entire study area the mean and median fire size dropped by 4.52 and 5.25 %, respectively, and the mean and median suppression expenditures by 6.57 and 5.03 %. Within treated areas the signal was much stronger, leading to 17.05 and 22.22 % reductions in mean and median wildfire size, respectively, and to 15.83 and 17.55 % reductions in mean and median suppression expenditures. Assessment of treatment efficacy is tied the spatial scale of analysis: as the size of analysis area increases so too does the proportion of fires that never interact with a fuel treatment, thus dampening the magnitude of the treatment effect. Here we adopt a conservative approach and assess all simulated fires within the entire fire modeling landscape.

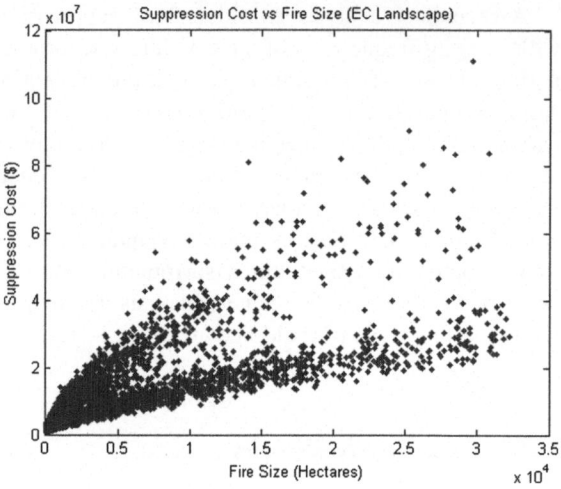

Fig. 5.4 Relationship between simulated fire size (x-axis) and suppression expenditures (y-axis), for escaped large wildland fires on the existing conditions (EC) landscape

In total 5,673 simulated wildfires grew to the large wildfire threshold (121 hectares) on either the EC or the PT landscape. In 396 cases, the ignition on PT landscape didn't grow sufficiently large and is assumed contained by initial attack efforts; in 6 of these instances the opposite occurred (the ignition grew to become large only on the EC landscape). Of the remaining 4,841 ignitions, 3,319 fires grew smaller on the PT landscape (average size reduction = 685 hectares), and 1,522 grew larger on the PT landscape (average size increase = 14 hectares). Thus, although the variability associated with spotting can result in different fire sizes on the EC and PT landscape for the same ignition, the magnitude of the treatment effect is quite apparent. Further, not only did fuel treatments tend to reduce the size of and expenditures on large wildfires, but treatments also reduced the likelihood of an ignition growing to become "large." Assuming equal initial attack effort on the EC and PT landscapes for all ignitions, the result is a net suppression expenditures savings in these instances.

Figure 5.5 presents changes in simulated annual burn probabilities across the EC and PT landscapes. Reductions in burn probability are evident, especially within treated areas and in areas proximal to fuel treatments (Fig. 5.3), due to reductions in rate of spread and final fire size. Burn probabilities do increase in some areas due to stochastic spotting, although the magnitude of these increases is small relative to reductions elsewhere on the landscape. Figure 5.6 presents a smoothed expenditure savings surface comparing expected suppression expenditures associated with each ignition location. The spatial pattern of expenditure savings is similar to patterns of burn probability reductions, with wildfires that ignited within or proximal to treated areas showing the greatest reductions.

Comparing the EC and PT landscapes on an annualized rather than per-fire basis captures those fire seasons in which no large fires occur and those fire seasons in which multiple large fires occur. Distributions of annual area burned and annual suppression expenditures can therefore evince high variability across

Fig. 5.5 Reductions in annual burn probability (calculated as the EC landscape values minus the PT landscape values, for burn probabilities at each pixel), across the fire modeling landscape. Figure from Thompson et al. (2013d)

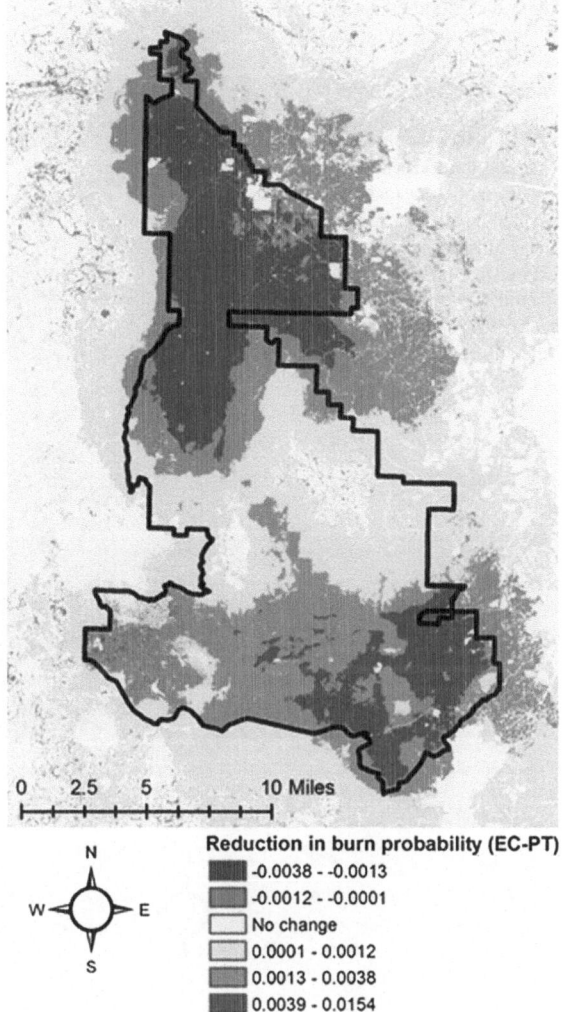

simulated fire seasons. At least one large wildfire occurred on either the EC or PT landscape on 3,629 out of the 10,000 simulated fire seasons. A total of 3,626 seasons on the EC landscape had at least one large wildfire, which dropped to 3,463 on the PT landscape, again confirming that after treatment some ignitions don't grow to become large (and further that in rare instances—in this case 3 seasons—small wildfires on the EC landscape can grow to exceed 121 hectares on the PT landscape). The largest number of wildfires to occur in a single simulated season on either landscape was 7, with the EC landscape averaging 0.57 and the PT landscape 0.53 large wildfire ignitions per season, a 6.88 % decrease. Mean annual area burned on the EC and PT landscapes were 2,184 and 1,942 hectares per season, respectively, an 11.09 % decrease. Lastly, mean annual suppression

Fig. 5.6 Smoothed surface
of suppression expenditure
savings (calculated as the EC
landscape values minus the
PT landscape values, for
expenditures associated with
each ignition location), across
the fire modeling landscape.
Results are presented in log
scale, illustrating vast
differences in the magnitude
of modeled expenditure
savings (*large*) and
expenditure increases (*small*)

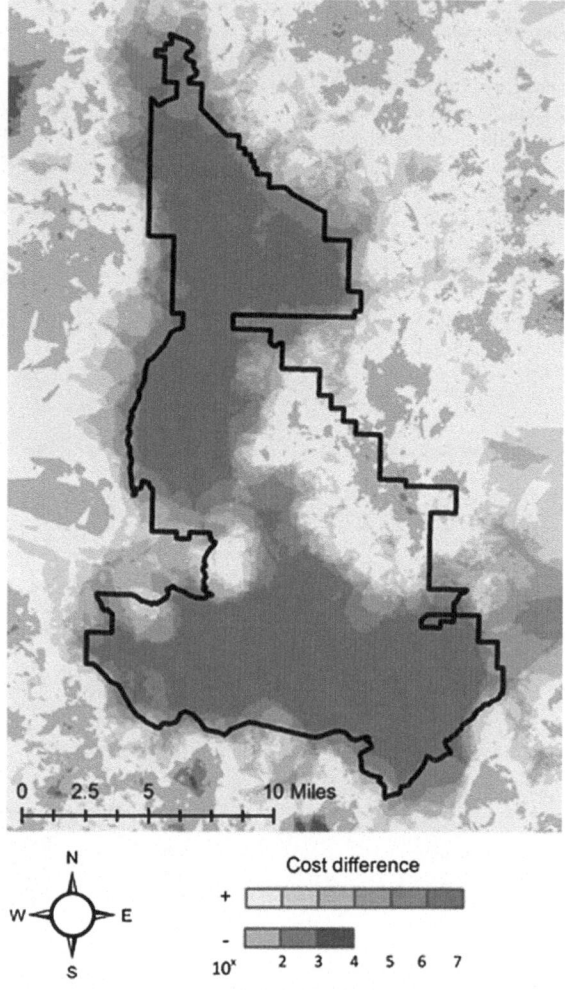

expenditures were \$5,094,727 and \$4,432,626 per season on the EC and PT landscapes, respectively, a 13.00 % decrease.

Figure 5.7 presents an empirical probability density function for the change in annual suppression expenditures between the EC and PT landscapes. The distribution is conditional on large wildfire occurrence (i.e., seasons with zero wildfires are excluded). The majority (2,547 of 3,629, or 70.18 %) of fire seasons show expenditure savings, averaging \$2,614,530. Most of these seasons show only small or moderate expenditure savings, however: the 25th percentile is \$27,897, the median is \$498,368, the 75th percentile is \$3,018,429, and the 90th percentile is \$ 7,850,444. A smaller fraction (856 of 3,629, or 23.59 %) of fire seasons actually lead to expenditure increases, which again is due to modest increases in fire sizes due to stochastic spotting. The magnitude of suppression expenditure increases is

Fig. 5.7 Empirical probability density function (PDF) for change in annual suppression expenditures (calculated as the EC landscape values minus the PT landscape values, for expenditures associated with each simulated wildfire season with at least one large wildfire)

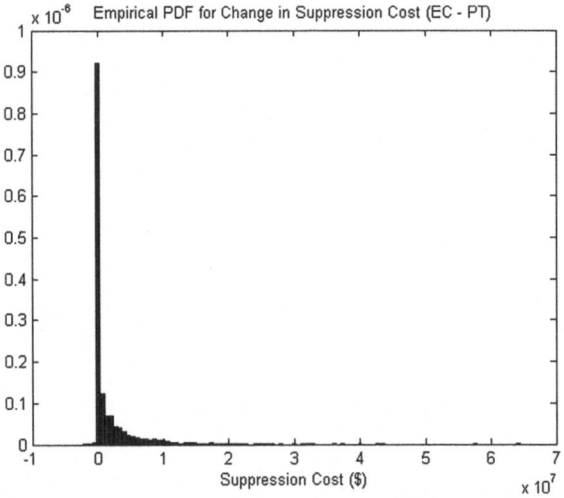

far smaller, with mean and median suppression expenditure increases of $44,629 and $11,936, respectively.

Analysis of annualized results allows for probabilistic statements about the likelihood of the fuel treatments influencing future suppression expenditures. Firstly, there is a 36.29 % chance of seeing any large wildfires across the landscape. Secondly, in seasons that do experience at least one wildfire, expenditure savings are realized 70.18 % of the time. There is therefore a 25.47 % annual chance of seeing suppression expenditure reductions of any magnitude. Referring to the percentile values presented in the above paragraph, there is a 12.75 % chance of saving at least $498,368, a 6.37 % chance of saving at least $2,220,084, and only a 2.55 % chance of saving at least $7,850,444.

The second management scenario explores the impact of increasing frequency of large wildfire occurrence on the EC and PT landscapes. These results are in effect annualized, but with a predetermined number of large wildfires per year rather than a simulated number of fires per year. Figure 5.8 presents empirical cumulative distribution functions for annual suppression expenditures on the EC and PT landscapes, over one, three, and five escaped large wildfires per year. As with Fig. 5.7, results are conditional on the number of large wildfires. On the EC landscape, the likelihood of seeing five large wildfires in a given season (0.29 %) is nearly two orders of magnitude lower than the likelihood of seeing one large wildfire (22.82 %), although these probabilities could change in the future. Horizontal distances between any two expenditure curves represent expected expenditure savings at a given percentile value. The curves shift to the right as the amount of wildfire increases, indicating higher expected annual suppression expenditures.

Moving between these 1–3–5 curves in Fig. 5.8 indicates the potential savings associated with investments in prevention and preparedness, or conversely

Fig. 5.8 Empirical
cumulative distribution
function (CDF) for annual
suppression, on both the EC
and PT fire modeling
landscapes, assuming 1, 3, or
5 large wildfires per fire
season

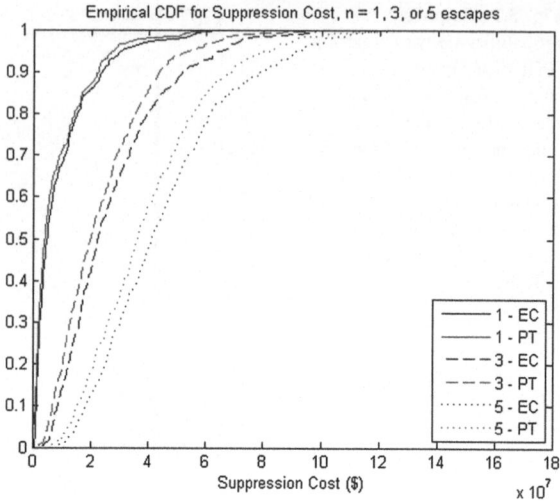

indicates potential expenditure increases associated with disinvestments or with a changing climate (we should note that a changing climate could also influence fire extent not just occurrence, a phenomenon we do not model). Moving between EC and PT curves for a single escape are fairly close together, but the horizontal distance between the EC and PT curves increases with an increasing number of wildfires per year. Increased wildfire on the landscape increases the likelihood of wildfires interacting with fuel treatments, and thus increases the expected efficacy of fuel treatments. An indirect but important implication of this result is that the likely efficacy of fuel treatments is greatest when treatments are placed in areas that are most likely to experience wildfire.

5.3 Discussion

The modeling techniques presented here illustrate an application of the SCI model paired with a wildfire simulation model for evaluating likely suppression expenditures on various landscapes, as well as for evaluating the likely expenditure impacts of various wildfire management and risk mitigation strategies. In turn this approach allows analysts to address a variety of salient wildfire management and policy questions, including comparative assessment of alternative wildfire management strategies, comparing expected suppression expenditures across landscapes and geographic areas, and possibly comparing expected suppression expenditures for direct protection areas and other agreements that span ownership boundaries. Fundamentally the modeling results can be used to help determine whether the up-front cost of fuel treatment activities are likely to be outweighed or at least partially offset by future, uncertain suppression expenditure savings.

Results will be dependent on the fire modeling techniques used, as well as the approach to quantifying marginal changes in wildfire activity associated with management actions, which will come with a degree of error and uncertainty. Sensitivity and scenario analyses can examine a range of potential suppression expenditure impacts given uncertainty surrounding treatment efficacy and future wildfire activity. Whether suppression expenditures actually change will depend largely on the decisions of fire managers, as well as other factors potentially influencing their decision-making process (Calkin et al. 2013; Donovan et al. 2011; Wibbenmeyer et al. 2013; Wilson et al. 2011). In the case of fuel treatments, whether incident managers are aware of the presence and likely efficacy of fuel treatments will in large part dictate potential changes in incident response, although fire behavior observations could still indirectly lead to different responses than might have occurred had the treatments not been implemented.

A critical variable to be incorporated into the analysis is the likelihood of a treatment ever interacting with fire. This variable will in turn drive the likelihood of measurable suppression expenditure savings. Hence the utility of stochastic wildfire simulation and burn probability modeling techniques are seeing increasing use across the wildfire management spectrum (Finney et al. 2011a; Bar Massada et al. 2009; Thompson et al. 2013c). The case study results on a fire-prone landscape indicated that landscape-scale fuel treatments could lead to reduced wildfire sizes and expected suppression expenditures, and yet the likelihood of seeing expenditure savings in any given fire season was just one in four.

Another important variable in the analysis is the spatial scale at which treatment effects are evaluated. The fire modeling landscape by necessity extends beyond the scale of treated areas alone, to account for off-site ignitions that can burn into treated areas and affect on-site burn probabilities. As described earlier, the larger the spatial scale, the greater the proportion of fires that may never impact fires, so treatment effects may appear dampened. One option is to record net savings in absolute rather than relative terms, which in effect ignores wildfires where there are no discernible treatment effects and is not as influenced by scale. Another option could be to delineate a "fireshed" for fuel treatments as the area within which simulated ignitions could reach treated areas (Thompson et al. 2013b). All wildfire ignitions within this fireshed would then be included in the assessment, regardless of whether or not they grew to interact with treated areas.

It is critical to align the analytical framework with the landscape context and the wildfire management objectives. A variety of modeling approaches exist, with varying levels of suitability for addressing specific questions. Fuel treatments, for instance, can have a wide variety of objectives, some of which may have little direct bearing on reducing suppression expenditures. The modeling technique illustrated here addressed expenditure impacts from changes to fire size, but not changes to fire intensity or burn severity. The current SCI model does not directly assess the expenditure impacts of fire intensity or burn severity, which would limit its use for such applications. Fitch et al. (2013) present an alternative framework that could help identify expenditure impacts where fuel treatments, or other management actions, lead to lower severity wildfires.

An additional limitation of the approach described here is that fire size is modeled independently of expenditures. That is, the modeling framework sequentially generates wildfire outcomes, then uses those outcomes to estimate expenditure outcomes. Effective wildfire incident management efforts are likely to alter the course of fires and potentially affect final fire size. Yoder and Gebert (2012) jointly model fire size and expenditures based on ignition point characteristics (like the SCI model does for expenditures alone). This approach accounts for the possibility that suppression effort can affect fire size, but does not capture how the spatial distribution of changes in fuel conditions may affect incident management efforts. Fire growth potential and final size are complex functions of topography, fuels, weather, and containment effort; changes to these factors are not well represented in an expenditures model based on ignition characteristics alone.

Future work could bring in a number of extensions and refinements. First, as introduced earlier, alternative expenditure modeling approaches could be more suitable or more appropriate contextually. Spatial expenditure models that better account for landscape characteristics, and mechanistic rather than statistical models that better account for firefighting resource usage over the duration of the wildfire could both lead to improved expenditure estimation. Second, expanded modeling of wildfire management activities, especially initial attack modeling and suppression response could be brought to bear. An ability to comprehensively model the consequences of investments across the wildfire management spectrum could better inform tradeoff analysis and lead to optimization approaches. Ultimately linking suppression expenditure models with wildfire simulation models could lead to an improved ability to understand and manage the financial impacts of large wildfires.

Chapter 6
Outlook and Future Research Directions for the Economics of Wildfire Management

Abstract This chapter summarizes the findings on the state-of-the-art for wildfire management economics research. The previous chapters have demonstrated that there is considerable complexity in describing the determinants of large fire suppression expenditures, and much remains to be known about the costs of fire management. Future research directions are summarized, including community responses to perceived fire risk, managerial incentives and risk preferences, and socio-political factors in fire management decisions. Current trends in large wildfire activity and management expenditures suggest that further economic research may be useful in improving the effectiveness and efficiency of wildfire management.

Keywords Wildfire suppression · Suppression expenditures · Cost effectiveness · Manager incentives · Risk perception · Risk management · Net value change

Wildfire activity and its associated damages and management expenditures have increased in recent years in the United States and throughout the world. Several factors account for these increases, including increased human development within fire prone areas, fuel buildup through land use change and past wildfire suppression, and increased frequency and severity of extreme fire weather due to global climate change.

This book has focused primarily on management expenditures for wildfires managed by the USFS—the governmental agency with the largest wildfire management responsibility in the United States. As the previous chapters demonstrate, there has been considerable progress toward understanding the factors that influence management expenditures for large wildfires. Yet uncertainties and difficulties remain due to the heterogeneous nature of wildfires themselves. Large wildfires show substantial variability in terms of geographic setting, fire behavior, values at risk, socio-political environment, and management response.

The fundamentals of wildfire economics were laid out almost a century ago when Sparhawk (1925) developed the Cost Plus Loss (CPL) model. CPL shows that the optimal wildfire management program minimizes management expenditures plus resource loss due to wildfire. Improvements to the Sparhawk model have

M. S. Hand et al., *Economics of Wildfire Management*, SpringerBriefs in Fire,
DOI: 10.1007/978-1-4939-0578-2_6, © The Author(s) 2014

been recommended, such as replacing CPL with Cost Plus Net Value Change (C + NVC) and the recognition that optimal budgets for pre-suppression, suppression, and Net Value Change cannot be simultaneously determined (Donovan and Rideout 2003). Yet the fundamental concept behind CPL and C + NVC remains sound; cost-effective investments in wildfire suppression are those that aim to minimize programmatic costs and wildfire-related net loss.

Wildfire expenditure models studied in this book (such as the SCI) are currently able to provide a wealth of information that can be used to evaluate programmatic costs in a CPL or C + NVC framework. Major findings described in the preceding chapters indicate the advances, potential uses, and limitations of expenditure models in this regard, including:

- Regression expenditure models are reasonably good at identifying primary factors related to management expenditures using variables readily available in Agency reporting systems, though considerable unexplained variation remains (Chap. 2).
- SCI-type models have become well established as a tool to examine management performance (on individual fires and for a fire season as a whole) and provide decision support during an incident. The models are best considered as coarse filters for identifying fires with above average expenditures, and there may be instances when higher than average expenditures cannot be explained by observable factors, such as potential risk or socio-political influences. In these cases, thoughtful review of management decision making may help reduce unnecessary expenditures in the future (Chap. 2).
- Aggregate regional expenditures over time follow an unpredictable path that does not adhere to a consistent time trend, and regional expenditures in the West tend to move together. This suggests that increases in aggregate expenditures over time are not inexorable and leaves open the possibility that observable factors, such as climate variations and the geographic distribution of fires in a given year, help determine year-over-year expenditures (Chap. 3).
- Differences in regional expenditure structures persist and are difficult to explain with observable data, whereas differences in expenditure structures between high- and low-cost years are explained by differences in observable fire characteristics. Results indicate that expenditure containment efforts may benefit from focusing on management activities in specific regions (i.e., California and the Northwest) and exploring the role of human factors (e.g., risk biases, social and political pressures) that are not captured in regression expenditure models (Chap. 3).
- Spatially descriptive fire characteristics are useful for explaining variations in fire expenditures in the western United States and can account for spatial relationships (i.e., autocorrelation) between observations. These models may be useful for applications where the spatial arrangement of landscape characteristics that affect fire behavior and expenditures may be important (e.g., analysis of fuel treatment costs and benefits) (Chap. 4).

- Expenditure models can be paired with fire simulation models to help quantify the expenditure consequences of land management activities (e.g., landscape-scale fuel treatment programs). In an example application, expenditure models are capable of describing how fuel treatments may affect future suppression expenditures when fuel treatments affect the expected size of future fires. This suggests that cost-effective planning of fuel treatment investments involve tradeoffs between treatment costs, changes in future suppression expenditures, and effects on public and private resources (Chap. 5).

In light of these general findings, the challenge for managers, policy makers, and researchers is to begin to fill in the gaps in knowledge about the determinants of wildfire management expenditures and how public agencies can plan wildfire management programs for more cost-effective outcomes. Potentially fruitful lines of research could consider how additional socio-political factors are related to suppression expenditures, such as expectations of management partners involved in fire management, community characteristics, and managerial experience and risk preferences.

Future research may benefit from engaging with emerging social science research that attempts to better understand how communities respond to perceived fire risk. Within the USFS, local line officers (e.g. District Rangers and Forest Supervisors) are responsible for developing fire management strategies and delegating the operational authority to the Incident Management Teams. Managerial incentives (Donovan and Brown 2005; Thompson et al. 2013a) and managerial biases and risk perception may be important in this process (Wilson et al. 2011; Wibbenmeyer et al. 2013), yet the role of such factors in decision making and strategic choices are only beginning to be understood in a wildfire management context.

Fundamentally, there is limited understanding of how fire suppression actions lead to the conclusion of fire events (Finney et al. 2009; Holmes and Calkin 2013). Within this book we have focused on the use of econometric regression models to understand wildfire suppression expenditures. Alternative models such as production theory may hold promise, but the range of large wildfire strategies makes it difficult to determine what essentially is being produced by teams managing a fire (Holmes and Calkin 2013).

This book has demonstrated that there is considerable variation in large fire suppression expenditures, and that there remains substantial uncertainty regarding many of the factors driving these expenditures. Yet estimating the net value change due to fire is even more challenging (Venn and Calkin 2011; Thompson and Calkin 2011). Given the current trends in large wildfire activity and the impacts of wildfire management on public agencies' budgets, economic research is greatly needed to improve the effectiveness and efficiency of wildfire management. Increased residential development within fire prone areas and the potential for increased wildfire activity due to climate change require a comprehensive examination of how different fire management investments in prevention, preparedness, fuel treatment and large fire suppression can best achieve societal goals.

References

Abatzoglou JT (2013) Development of gridded surface meteorological data for ecological applications and modelling. Int J Climatol 33(1):121–131. doi:10.1002/joc.3413

Abt KL, Prestemon JP, Gebert KM (2009) Wildfire suppression cost forecasts for the US Forest Service. J Forest 107(4):173–178

Ager AA, Vaillant NM, Finney MA (2010) A comparison of landscape fuel treatment strategies to mitigate wildland fire risk in the urban interface and preserve old forest structure. Forest Ecol Manage 259(8):1556–1570. doi:10.1016/j.foreco.2010.01.032

Ager AA, Vaillant NM, McMahan A (2013) Restoration of fire in managed forests: a model to prioritize landscapes and analyze tradeoffs. Ecosphere 4(2):19. doi:10.1890/ES13-00007.1 (article 29)

Bar Massada A, Radeloff VC, Stewart SI, Hawbaker TJ (2009) Wildfire risk in the wildland-urban interface: a simulation study in northwestern Wisconsin. Forest Ecol Manage 258(9):1990–1999. doi:10.1016/j.foreco.2009.07.051

Bessie WC, Johnson EA (1995) The relative importance of fuels and weather on fire behavior in subalpine forests. Ecology 76(3):747–762. doi:10.2307/1939341

Bjørnstad ON, Falck W (2001) Nonparametric spatial covariance functions: estimation and testing. Environ Ecol Statistics 8(1):53–70

Botti SJ (1999) The National Park Service wildland fire management program. In: Proceedings of the symposium on fire economics, planning, and policy: bottom lines, Pacific Southwest Research Station, San Diego, CA, April 1999, pp 7–13

Butry DT, Mercer E, Prestemon JP, Pye JM, Holmes TP (2001) What is the price of catastrophic wildfire? J Forest 99(11):9–17

Butry DT, Prestemon JP, Abt KL, Sutphen R (2010) Economic optimisation of wildfire intervention activities. Int J Wildland Fire 19(5):659–672. doi:10.1071/WF09090

Calkin DE, Ager AA, Thompson MP, Finney MA, Lee DC, Quigley TM, McHugh CW (2011a) A comparative risk assessment framework for wildland fire management: the 2010 cohesive strategy science report. RMRS-GTR-262, U.S. Department of Agriculture, Forest Service, Rocky Mountain Research Station, p 63

Calkin DE, Finney MA, Ager AA, Thompson MP, Gebert KM (2011) Progress towards and barriers to implementation of a risk framework for US federal wildland fire policy and decision making. Forest Policy Econ 13(5):378–389. doi:10.1016/j.forpol.2011.02.007

Calkin DE, Gebert KM, Jones G, Neilson RP (2005) Forest service large fire area burned and suppression expenditure trends 1970-2002. J Forest 103(4):179–183

Calkin DE, Rieck JD, Hyde KD, Kaiden JD (2011) Built structure identification in wildland fire decision support. Int J Wildland Fire 20(1):78–90. doi:10.1071/WF09137

Calkin DE, Thompson MP, Finney MA, Hyde KD (2011) A real-time risk assessment tool supporting wildland fire decisionmaking. J Forest 109(5):274–280

Calkin DE, Venn TJ, Wibbenmeyer MJ, Thompson MP (2013) Estimating US federal wildland fire managers' preferences toward competing strategic suppression objectives. Int J Wildland Fire 22(2):212–222. doi:10.1071/WF11075

M. S. Hand et al., *Economics of Wildfire Management*, SpringerBriefs in Fire, DOI: 10.1007/978-1-4939-0578-2, © The Author(s) 2014

Canton-Thompson J, Gebert KM, Thompson B, Jones G, Calkin DE, Donovan GH (2008) External human factors in incident management team decisionmaking and their effect on large fire suppression expenditures. J Forest 106(8):416–424

Cardille JA, Ventura SJ, Turner MG (2001) Environmental and social factors influencing wildfires in the Upper Midwest, United States. Ecol Appl 11(1):111–127. doi:10.2307/3061060

Chevan A, Sutherland M (1991) Hierarchical partitioning. Am Statistician 45(2):90–96. doi:10.1080/00031305.1991.10475776

Cochrane MA, Moran CJ, Wimberly MC, Baer AD, Finney MA, Beckendorf KL, Eidenshink J, Zhu Z (2012) Estimation of wildfire size and risk changes due to fuels treatments. Int J Wildland Fire 21(4):357–367. doi:10.1071/WF11079

Collins BM, Omi PN, Chapman PL (2006) Regional relationships between climate and wildfire-burned area in the Interior West, USA. Can J Forest Res 36(3):699–709. doi:10.1139/X05-264

Collins BM, Stephens SL, Moghaddas JJ, Battles J (2010) Challenges and approaches in planning fuel treatments across fire-excluded forested landscapes. J Forest 108(1):24–31

Crimmins MA, Comrie AC (2004) Interactions between antecedent climate and wildfire variability across south-eastern Arizona. Int J Wildland Fire 13(4):455–466. doi:10.1071/WF03064

DeJong DN, Nankervis JC, Savin NE, Whiteman CH (1992) The power problems of unit root tests in time series with autoregressive errors. J Economet 53(1–3):323–343

Deeming JE, Burgan RE, Cohen JD (1977) The National Fire-Danger Rating System—1978. General Technical Report, U.S. Department of Agriculture, Forest Service, Ogden, UT

Donovan GH, Brown TC (2005) An alternative incentive structure for wildfire management on national forest land. Forest Sci 51(5):387–395

Donovan GH, Prestemon JP, Gebert KM (2011) The effect of newspaper coverage and political pressure on wildfire suppression costs. Soc Nat Resources 24(8):785–798. doi:10.1080/08941921003649482

Donovan GH, Rideout DB (2003) A reformulation of the cost plus net value change (C+NVC) model of wildfire economics. Forest Sci 49(2):318–323

Efron B, Tibshirani RJ (1993) An introduction to the bootstrap, vol 1, 1st edn. Chapman and Hall/CRC Press, Boca Raton

Enders W (2008) Applied econometric time series. Wiley, New York

Engle RF, Granger CWJ (1987) Co-integration and error correction: representation, estimation, and testing. Econometrica 55(2):251–276

Finney MA (2004) FARSITE: fire area simulator-model development and evaluation. RMRS-RP-4, U.S. Department of Agriculture, Forest Service, Rocky Mountain Research Station, p 47

Finney MA, Grenfell IC, McHugh CW (2009) Modeling containment of large wildfires using generalized linear mixed model analysis. Forest Sci 55(3):249–255

Finney MA, Grenfell IC, McHugh CW, Seli RC, Trethewey D, Stratton RD, Brittain S (2011) A method for ensemble wildland fire simulation. Environ Model Assess 16:153–167. doi:10.1007/s10666-010-9241-3

Finney MA, McHugh CW, Grenfell IC, Riley KL, Short KC (2011) A simulation of probabilistic wildfire risk components for the continental United States. Stochast Environ Res Risk Assess 25:973–1000. doi:10.1007/s00477-011-0462-z

Fire Program Analysis (2010) Fire program analysis charter. http://www.forestsandrangelands.gov/FPA/documents/overview/FPA_Charter_20101014.pdf. Accessed 11 Sept 2013

Fire Program Analysis (2012) Product management and scope statement. http://www.forestsandrangelands.gov/FPA/documents/overview/FPA_Scope_2012April.pdf. Accessed 11 Sept 2013

Fitch RA, Kim Y-S, Waltz AEM (2013) Forest restoration treatments: their effect on wildland fire suppression costs. http://library.eri.nau.edu/gsdl/collect/erilibra/index/assoc/D2013009.dir/doc.pdf. Accessed 15 May 2013

Flannigan MD, Stocks BJ, Wotton BM (2000) Climate change and forest fires. Sci Total Environ 262(3):221–229. doi:10.1016/S0048-9697(00)00524-6

Fried JS, Gilless JK, Spero J (2006) Analysing initial attack on wildland fires using stochastic simulation. Int J Wildland Fire 15:137–146. doi:10.1071/WF05027

Gebert KM (2007) Development of Department of Interior "stratified cost index" model: feasibility report. Unpublished report to the Department of Interior, Office of Wildland Fire Coordination, p 11

Gebert KM, Black AE (2012) Effect of suppression strategies on federal wildland fire expenditures. J Forest 110(2):65–73

Gebert KM, Calkin DE, Yoder J (2007) Estimating suppression expenditures for individual large wildland fires. West J Appl Forest 22(3):188–196

Gedalof ZM, Peterson DL, Mantua NJ (2005) Atmospheric, climatic, and ecological controls on extreme wildfire years in the northwestern United States. Ecol Appl 15(1):154–174. doi:10.1890/03-5116

Gill AM, Stephens SL (2009) Scientific and social challenges for the management of fire-prone wildland–urban interfaces. Environ Res Lett 4(3):034014. doi:10.1088/1748-9326/4/3/034014

Gonzalez-Caban A (1984) Costs of firefighting mopup activities. Research Note PSW-367. USDA Forest Service Pacific Southwest Forest and Range Experiment Station, p 5

Goodwin BK, Piggott NE (2001) Spatial market integration in the presence of threshold effects. Am J Agric Econ 83(2):302–317

Goodwin BK, Schroeder TC (1991) Cointegration tests and spatial price linkages in regional cattle markets. Am J Agric Econ 73(2):452–464. doi:10.2307/1242730

Gorte JK, Gorte RW (1979) Application of economic techniques to fire management - a status review and evaluation. INT-GTR-53. U.S. Department of Agriculture, Forest Service, Intermountain Research Station, p 26

Greene WH (2003) Econometric analysis, vol 1, 5th edn. Prentice Hall, Upper Saddle River

Gude PH, Jones K, Rasker R, Greenwood MC (2013) Evidence for the effect of homes on wildfire suppression costs. Int J Wildland Fire 22(4):537–548. doi:10.1071/WF11095

Hesseln H, Amacher GS, Deskins A (2010) Economic analysis of geospatial technologies for wildfire suppression. Int J Wildland Fire 19(4):468–477. doi:10.1071/WF08155

Holmes TP, Calkin DE (2013) Econometric analysis of fire suppression production functions for large wildland fires. Int J Wildland Fire 22:246–255. doi:10.1071/WF11098

Houtman RM, Montgomery CA, Gagnon AR, Calkin DE, Dietterich TG, McGregor S, Crowley M (2013) Allowing a wildfire to burn: estimating the effect on future fire suppression costs. Int J Wildland Fire 22(7):871–882. doi:10.1071/WF12157

Kochi I, Champ PA, Loomis JB, Donovan GH (2012) Valuing mortality impacts of smoke exposure from major southern California wildfires. J Forest Econ 18(1):61–75. doi:10.1016/j.jfe.2011.10.002

Liang J, Calkin DE, Gebert KM, Venn TJ, Silverstein RP (2008) Factors influencing large wildland fire suppression expenditures. Int J Wildland Fire 17(5):650–659. doi:10.1071/WF07010

Lundgren S (1999)The national fire management analysis system (NFMAS) past 2000: a new horizon. In: Gonzalez-Caban A, Omi PN (eds) Proceedings of the symposium on fire economics, planning, and policy: bottom lines. USDA Forest Service General Technical Report PSW-173, Pacific Southwest Research Station, San Diego, CA, 5–9 April 1999, pp 71–78

McLeod AI, Yu H, Mahdi E (2011) Time series analysis with R. In: Rao TS, Rao CR (eds) Handbook of statistics, vol 30, 1st edn. Elsevier, Oxford, p 75

Mills TJ, Bratten FW (1982) FEES: design of a fire economics evaluation system. GTR-PSW-065, USDA Forest Service Pacific Southwest Forest and Range Experiment Station, p 26

Mozumder P, Helton R, Berrens RP (2009) Provision of a wildfire risk map: informing residents in the wildland urban interface. Risk Anal 29(11):1588–1600. doi:10.1111/j.1539-6924.2009.01289.x

Noonan-Wright E, Opperman TS, Finney MA, Zimmerman T, Seli RC, Elenze LM, Calkin DE, Fiedler JR (2011) Developing the U.S. wildland fire decision support system. J Combust 1–14. doi:10.1155/2011/168473 (Article ID 168473)

OIG (2004) Audit report: implementation of the healthy forests initiative. 08601-6-AT, U.S. Department of Agriculture, Office of Inspector General, Southeast Region, p 28

Pence M, Zimmerman T (2011) The wildland fire decision support system: integrating science, technology, and fire management. Fire Manage Today 71(1):18–22

Petrovic N, Carlson JM (2012) A decision-making framework for wildfire suppression. Int J Wildland Fire 21(8):927–937. doi:10.1071/WF11140

Pfaff B (2008) Analysis of integrated and cointegrated time series with R, vol 1, 2nd edn. Springer, New York

Pfeifer PE, Deutsch SJ (1980) A three-stage iterative procedure for space-time modeling. Technometrics 22(1):35–47. doi:10.1080/00401706.1980.10486099

Preisler HK, Westerling AL, Gebert KM, Munoz-Arriola F, Holmes TP (2011) Spatially explicit forecasts of large wildland fire probability and suppression costs for California. Int J Wildland Fire 20(4):508–517. doi:10.1071/WF09087

Prestemon JP, Abt KL, Gebert KM (2008) Suppression cost forecasts in advance of wildfire seasons. Forest Sci 54(4):381–396

Prestemon JP, Ham C, Abt KL (2012) Federal land assistance, management, and enhancement (FLAME) Act suppression expenditures for interior and agriculture agencies: March 2012 forecasts for Fiscal Year 2012. USDA Forest Service, Southern Research Station, p 14

Core Team R (2013) R: a language and environment for statistical computing. R Foundation for Statistical Computing, Vienna

Reinhardt ED, Keane RE, Calkin DE, Cohen JD (2008) Objectives and considerations for wildland fuel treatment in forested ecosystems of the interior western United States. Forest Ecol Manage 256(12):1997–2006. doi:10.1016/j.foreco.2008.09.016

Rideout DB, Ziesler PS (2008) Three great myths of wildland fire management. In: Gonzalez-Caban A (ed) Proceedings of the second international symposium on fire economics, planning, and policy: a global view, Cordoba, Spain, 19–22 April 2008, USDA Forest Service General Technical Report PSW-208, Pacific Southwest Research Station, p 7

Schoennagel TL, Veblen TT, Romme WH (2004) The interaction of fire, fuels, and climate across Rocky Mountain forests. BioScience 54(7):661–676. doi:10.1641/0006-3568(2004)054[0661:TIOFFA]2.0.CO;2

Scott J, Helmbrecht D, Thompson MP, Calkin DE, Marcille K (2012) Probabilistic assessment of wildfire hazard and municipal watershed exposure. Nat Hazards 64(1):707–728

Sparhawk WR (1925) The use of liability rating in planning forest fire protection. J Agric Res 30:693–762

Steele TW, Stier JC (1998) An economic evaluation of public and organized wildfire detection in Wisconsin. Int J Wildland Fire 8(4):205–215. doi:10.1071/WF9980205

Thompson MP, Calkin DE (2011) Uncertainty and risk in wildland fire management: a review. J Environ Manage 92(8):1895–1909. doi:10.1016/j.jenvman.2011.03.015

Thompson MP, Calkin DE, Finney MA, Gebert KM, Hand MS (2013) A risk-based approach to wildland fire budgetary planning. Forest Sci 59(1):63–77

Thompson MP, Scott J, Kaiden JD, Gilbertson-Day JW (2013) A polygon-based modeling approach to assess exposure of resources and assets to wildfire. Nat Hazards 67(2):627–644. doi:10.1007/s11069-013-0593-2

Thompson MP, Scott JH, Helmbrecht D, Calkin DE (2013) Integrated wildfire risk assessment: framework development and application on the Lewis and Clark National Forest in Montana, USA. Integr Environ Assess Manage 9(2):392–342

Thompson MP, Vaillant NM, Haas JR, Gebert KM, Stockmann KD (2013) Quantifying the potential impacts of fuel treatments on wildfire suppression costs. J Forest 111(1):49–58. doi:10.5849/jof.12-027

USDA Forest Service (2012) Future of America's forest and rangelands: forest service 2010 resources planning act assessment. General Technical Report WO-87, U.S. Department of Agriculture, Forest Service, p 198

USDA Forest Service, USDI Bureau of Land Management, and National Association of State Foresters (2003) Large fire cost reduction action plan. Federal Fire and Aviation Leadership Council. Accessed 15 Nov 2006

Venn TJ, Calkin DE (2011) Accommodating non-market values in evaluation of wildfire management in the United States: challenges and opportunities. Int J Wildland Fire 20(3):327–339. doi:10.1071/WF09095

Walsh C, Mac Nally R, Walsh MC (2003) The hier part package. Hierarchical partitioning R project for statistical computing URL: http://cranr-project.org

Westerling AL, Gershunov A, Brown TJ, Cayan DR, Dettinger MD (2003) Climate and wildfire in the Western United States. Bull Am Meteorol Soc 84(5):595–604. doi:10.1175/BAMS-84-5-595

Westerling AL, Gershunov A, Cayan DR, Barnett TP (2002) Long lead statistical forecasts of area burned in western U.S. wildfires by ecosystem province. Int J Wildland Fire 11(3/4):257–266. doi:10.1071/WF02009

Westerling AL, Hidalgo HG, Cayan DR, Swetnam TW (2006) Warming and earlier spring increase Western U.S. forest wildfire activity. Science 313(5789):940–943. doi:10.1126/science.1128834

Wibbenmeyer MJ, Hand MS, Calkin DE, Venn TJ, Thompson MP (2013) Risk preferences in strategic wildfire decision making: a choice experiment with U.S. wildfire managers. Risk Anal 33(6):1021–1037. doi:10.1111/j.1539-6924.2012.01894.x

Wilson RS, Winter PL, Maguire LA, Ascher T (2011) Managing wildfire events: risk-based decision making among a group of federal fire managers. Risk Anal 31(5):805–818

Yoder J, Gebert KM (2012) An econometric model for ex ante prediction of wildfire suppression costs. J Forest Econ 18(1):76–89. doi:10.1016/j.jfe.2011.10.003